SpringerBriefs in Applied Sciences and Technology

For further volumes:
http://www.springer.com/series/8884

Alberto Gemelli · Adriano Mancini
Claudia Diamantini · Sauro Longhi

GIS to Support Cost-Effective Decisions on Renewable Sources

Applications for Low Temperature Geothermal Energy

 Springer

Alberto Gemelli
Adriano Mancini
Ancona
Italy

Claudia Diamantini
Sauro Longhi
Università Politecnica delle Marche
Ancona
Italy

ISSN 2191-530X ISSN 2191-5318 (electronic)
ISBN 978-1-4471-5054-1 ISBN 978-1-4471-5055-8 (eBook)
DOI 10.1007/978-1-4471-5055-8
Springer London Heidelberg New York Dordrecht

Library of Congress Control Number: 2013934006

Printed on acid-free paper

Springer is part of Springer Science+Business Media (www.springer.com)

Contents

Abbreviations

ATI	Apparent Thermal Inertia
ASTER	Advanced Spaceborn Thermal Emission and Reflection Radiometer
AVHRR	Advanced Very High Resolution Radiometer
BHE	Borehole Heat Exchanger
BVQFE	Bayesian Vector Quantizer Feature Extraction
$C_{GSHP/front}$	Up-front cost of GSHP plant
$C_{GSHP/year}$	Yearly maintenance costs of the GSHP plant
$C_{GSHP\&SHW/front}$	Up-front cost of GSHP&SHW plant
$C_{GSHP\&SHW/year}$	Yearly maintenance costs of the GSHP&SHW plant
$C_{METsplit/front}$	Up-front cost of METsplit plant
$C_{METsplit/year}$	Yearly maintenance costs of the METsplit plant
$C_{MET\&DHW/front}$	Up-front cost of MET&DHW plant
$C_{MET\&DHW/year}$	Yearly maintenance costs of the MET&DHW plant
$C_{SHW/front}$	Up-front cost of SHW plant
$C_{Split/front}$	Up-front cost of Split plant
$C_{Split/year}$	Yearly maintenance costs of the Split plant
$C_{Split\&DHW/front}$	Up-front cost of Split and DHW plant
$C_{Split\&DHW/year}$	Yearly maintenance costs of the Split&DHW plant
COP	Coefficient of Performance
COP_{split}	Coefficient of Performance of Split plant
CORINE	Coordination of Information on the Environment
$CO_2Equiv_{GSHP/year}$	CO_2 equivalent emitted by GSHP plant yearly
$CO_2Equiv_{GSHP\&SHW/year}$	CO_2 equivalent emitted by GSHP&SHW plant
$CO_2Equiv_{METsplit/year}$	CO_2 equivalent emitted by METsplit plant yearly
$CO_2Equiv_{MET\&DHW/year}$	CO_2 equivalent emitted by MET&DHW plant yearly
$CO_2Equiv_{Split/year}$	CO_2 equivalent emitted by Split plant, yearly
$CO_2Equiv_{Split\&DHW/year}$	CO_2 equivalent emitted by Split&DHW plant yearly
DBFE	Decision Border Feature Extraction
DD	Degree-Day
DHW	Domestic Hot Water
DSS	Decision Support System
DTM	Digital Terrain Model

E	Yearly energy needs for coling a 100 m^2 class E house
E_{BHE}	Yearly energy for cooling extracted by BHE
E_{EL}	Electric consume by GSHP for heating, yearly
EDBFM	Effective Decision Border Feature Matrix
EER	Energy Efficiency Ratio
EER$_{split}$	Energy Efficiency Ratio of split subsystem
EF	Emission Factor
EF$_{EL}$	Emission Factor of Grid Electricity
EF$_{ME}$	Emission Factor of Methane
EPSC	Energy Service Provider Company
ESCO	Energy Service Company
FE	Feature Extraction
FOSS	Free Open Source Software
FS	Feature Selection
GIS	Geographic Information System
GOFR	Goal Oriented Feature Ranking
GSHP	Ground Source Heat Pump
GSHP&SHW	Integrated GSHP and Solar Hot Water plant
GST	Ground Surface Temperature
H_u	Yearly energy needs for heating a square meter of class E house
H	Yearly energy needs for heating a 100 m^2 class E house
H_{BHE}	Yearly energy for heating extracted by BHE
H_{DHW}	Yearly energy requirements for the DHW
H_{DHW+}	Yearly energy requirements for the DHW plus 50 % of H
H_{EL}	Electric consume by GSHP for heating, yearly
LDA	Linear Discriminant Analysis
LST	Land Surface Temperature
LTGE	Low Temperature Geothermal Energy
MCSA	Multi-Criteria Decision Analysis
METsplit	Hybrid plant composed of a methane subsystem for heating and a split subsystem for cooling
MET&DHW	Integrated system, a methane subsystem for thermo-regulation and DHW, and a Split subsystem for air cooling
MODIS	Moderate Resolution Imaging Spectroradiometer
MRT	MODIS Reprojection Tool
NIR	Near Infra-Red
OLDA	Orthogonal Linear Discriminant Analysis
P_{BHE}	Power demand to BHE
SAGA	System for Automated Geoscientific Analyses
SAT	Surface Air Temperature
sHE	Specific Heat Extraction

sHE_{agg}	Specific Heat Extraction aggregate
SHW	Plant of Solar Hot Water
Split	Plant composed of a split system for heating and cooling
Split&DHW	An integrated plant with a methane subsystem for DHW&Split subsystem for thermoregulation.
$Z_{BHE'}$	Depth of the BHE

Part I
Preliminary Study

Chapter 1
Decision Environment of Renewable Energy: The Case of Geothermal Energy

Abstract The exploitation of low temperature geothermal energy (LTGE) for thermoregulation is an expanding activity with various applications. The production of LTGE, although it is an efficient process, it has a cost benefit varying from one site to another depending, but not only, of natural factors. For local administrations and for those who invest in the provision of energy services, a regional model of the distribution of the LTGE resource is necessary for planning production, incentives, and investment. Also, the factors that influence the cost benefit of the resource must be studied in the spatial dimension. The construction of spatial models is a process requiring the acquisition of large amounts of data, the use of computer technology, and a substantial process design effort. In this chapter, the emphasis is placed on the support of Geographic Information System (GIS) in spatial modeling of LTGE cost benefit. The techniques for collecting spatial data on a large geographic scale are introduced. The technical and organizational aspects discussed delineate an information environment aimed to providing decision support in the regional development of LTGE.

1.1 Introduction

The term *renewable energy* usually refers to energy produced with a positive net balance between production and consumption. Low Temperature Geothermal Energy (LTGE) is a notable example of renewable energy that is mainly suitable for solving the problem of domestic thermoregulation. LTGE is theoretically available anywhere on the Earth's surface and has the potential to play a key role in minimizing polluting emissions. In conjunction with other renewable energy sources, LTGE makes significant contributions to the new economic model referred to as the green economy. The concept of the green economy involves the sensitivity of citizens to the problems of pollution, governments' strategies to reduce energy dependence, and the economic attractiveness of investment in the "green economy" to private companies. Each stakeholder involved in renewable energy decides on its actions based on pros–cons analysis. These types of analyses require a large, complex, and heterogeneous body of information placed in context in the area where the renewable energy project is proposed. The need for accurate information is mainly due to long-term return on investment (ROI) considerations, with investment in the green economy requiring a period of several years to produce a profit, with a longer period than other alternatives. The main pro of this investment is the limited risk, and the fact that the profit can be accurately esti-mated. The pros–cons analysis of renewable energy cannot be the same for a whole region, making it necessary to conduct analyses for small subregions con-sidering different "boundary" conditions. Two main factors influence the analysis: (a) the geospatial variability of energy potential (production/consumption); (b) distributed production, where medium–small plants are considered (e.g., photovoltaic energy). Geographic information systems (GIS) represent a techno-logical tool capable of performing such an accurate evaluation of renewable energy for a given region, regardless of size.

In the past, the analysis of the production and consumption of energy resources, also known as *energy planning*, has been conducted without considering the geospatial relationship as a dominant factor. Often, plans have been publicly released in the form of simple general energy budgets, mainly for historical rea-sons due to (a) energy production from standard fossil fuels and (b) energy transportation to remote sites of consumption. In contrast, in a distributed approach to energy production, an energy plan requires a detailed and accurate study of available local resources. GIS provides a holistic view of complex situations at all scales (temporal, spatial) and allows for the cohesive analysis of a large set of factors behind the decision problem. GIS is basically an integration technology based on a combination of hardware, software, data and procedures for the col-lection, storage, manipulation, analysis and display of spatially distributed infor-mation on the phenomena under study to support a *decision process*.

In renewable energy planning, GIS enables the collection, integration, and processing of georeferenced socioeconomic data and features that characterize the geothermal resources. The variables that influence the planning are virtually endless, as are the alternative technologies available for resource exploitation. Therefore, specific strategies are implemented in the GIS for the selection and processing of the most significant information.

This book describes a GIS-based process for the systematic evaluation of geothermal resources across a wide region. The main technical aim of this work is to construct spatial decision-making models. The procedure developed here is innovative and easy to apply to a new case study area. The procedure starts by gathering natural and economic data and, through an innovative combination of methods for automatic reasoning, produces a regional model of the cost-effec-tiveness of geothermal production. Furthermore, a set of alternative plants are evaluated at different sites within the territory, and the features of their con-struction and performance are evaluated to decide which project solution is the most convenient, also considering the expectations of different stakeholders such as homeowners and investors. This process is tested in a specific case study, providing a real/practical example of its applicability. Unlike other renewable energy sources, geothermal energy requires knowledge of underground geological features. A notable feature of the method proposed here is the processing of freely available remote sensed data to construct a geological model of a wide region.

1.2 Stakeholders of Geothermal Energy

An energy resource is energy extracted from the environment to meet a wide set of human needs. The concept of a resource is determined by the amount and usefulness of the resource in relation to the interests of the stakeholders that exploit that resource. This section analyzes the expectations around geothermal resources.

The exploitation of LTGE is attracting the interest of many "entities" including consumers (householders), companies investing in civil/industrial plants, and public administrations. LTGE has a large set of potential applications, including the provision of temperature control for greenhouses, the drying of grains, and the provision of district-wide heating. LTGE also has a positive impact on the energy supply of the community, the reduction in polluting emissions, and economically sustainable development. During the last ten years, the thermal power generated by LTGE has increased by 12.3 % per year and 50 % of the production is dedicated to domestic thermoregulation (Lund et al. 2010; Bertani 2012). The power generated by LTGE is forecast to increase by a factor of 10x by 2050. Data updated in 2010 reveal that an average LTGE plant has a power of 5 kWt and that geothermal systems for domestic heating installed in Germany and Italy, countries with similar heating demand, have total capacities of 2,230 and 92 MWt, respectively. In the rest of the world, the countries extracting the greatest total amount of geothermal power are the USA (12.611 MWt) and China (8.898 MWt), followed by the lightly populated Sweden (4.460 MWt). These numbers demonstrate the youth of the geothermal market.

The stakeholder analysis for renewable energies is a complex process as described in Sudhakar and Painuly (2004). Focusing on the regional exploitation of the LTGE, the main stakeholders are as follows:

- *Users*: householders;
- *Investors*: Energy Services Companies (ESCOs);
- *Public Administrations*: local/regional government.

Each stakeholder category has specific expectations from the resource and bases its decisions on indicators that provide real criteria on which to base decisions and actions to execute (governance). Table 1.1 analyzes the stakeholders with respect to their expectations, indicators, and governance policies. Each stakeholder

Table 1.1 Stakeholder analysis

Stakeholder	Expectations	Indicators	Governance
Users	Saving on bill	Cost for the plants, break-even point period	Selection of the most convenient solution
	Energy requalification	Energy classes	Energy performance certificate
Investors	Financial investments: safe and profitable	Econometric	Asset immobilization
	Profitability	Market size	Geomarketing studies
Public administrations	Energy autonomy	Energy balance	Energy planning
	Emission reduction	Change of equivalent CO_2	Sensitization campaign
	Public accessibility to the resource	Natural potential	Study the regional area

category satisfies its expectations by choosing an action from a set of feasible solutions using the set of indicators.

The regional development of the resource is (must be) sponsored or pulled by the public administration. This entity has to solve the most complex problems as the policies affect all stakeholders. The goal of the public administration managing the economic/environmental development plan is to make the right choices and define standards (Haehnlein et al. 2010). Another important aspect for the government is to consider the ecological implications. Geothermal technology has a negligible impact on emissions because the physical process of energy generation does not involve a combustion process (Blum et al. 2010). Thus, the development of geothermal energy could reduce CO_2 emissions by replacing fossil energy sources.

The Energy Service Company (ESCO) (Vine 2005) is a typical example of a corporate stakeholder involved in the exploitation of LTGE. The mission of ES-COs is to create a form of financial investment by supporting householders to improve the energy efficiency and participating on the profits due to savings. Another category of company is the Energy Service Provider Company (ESPC). These companies usually offer a wide set of services such as supply, installation, and maintenance. The success of ESPCs depends on the demand for geothermal plants and on the appropriate selection of an area where the geothermal potential guarantees a medium- to short-term break-even point.

An information system (regional) is essential in helping the government to (a) define the current state of the resource, (b) predict its long-term evolution, and (c) lead the development based on the interests of the community (considering the ROI). For investors, an information system provides the knowledge necessary to decide on investment strategies and cope with the rapid and unexpected changes in this environment. GIS presents an opportunity for companies to obtain complete knowledge of the user-base and the available energy resources (potential sources of earnings).

1.3 Modeling the Ground as a Storage Reservoir of Thermal Energy

This chapter analyzes the factors that determine the natural potential of geothermal energy. The scientific literature on geothermal energy (Dickson and Fanelli 2005) contains a dichotomy:

- "Low Temperature Geothermal Energy" (LTGE) or *Low Enthalpy Geothermal Energy* is present at temperatures lower than 100 °C and mainly results from the accumulation of solar energy (irradiation) in the soil. The net power is estimated at 10–20 W/m². LTGE is available everywhere, although some areas are more suitable for the exploitation of this source of energy. LTGE can be used directly by extracting heat from the ground during the winter season or "pumping it in" during the summer season. The net thermal power roughly coincides with the average energy demand for a standard house.
- Classic geothermal energy refers to the generation of high temperatures by heat sources inside the Earth and is only available in rare sites (e.g., a famous site in Italy is Larderello). This resource cannot be used directly, but rather is used to produce electric energy. Note that the contribution of the internal heat source to the LTGE potential is limited (<0.1 W/m²) compared with the contribution from solar irradiation.

This book focuses on low temperature geothermal energy (LTGE).

The rock in the subsoil can be considered a large heat storage reservoir. Up to a few meters depth, the temperature of the rocks varies daily following the day/night cycles of the sun, but at greater depths the amount of heat is stable, and the temperature remains at a value approximately equal to the annual average of the *ground surface temperatures*. The rocks are thermally described by the *specific heat capacity* which is the capability of a given type of rock to store heat. It is usually expressed in Joule per Kelvin per kilogram ($J \cdot K^{-1} \cdot kg^{-1}$). If multiplied for the *specific weight,* it gives the *volumetric heat capacity* ($J \cdot K^{-1} \cdot m^{-3}$). The *thermal conductivity* ($W \cdot m^{-1} \cdot K^{-1}$) describes the amount of heat that spreads through the rock due to thermal conduction. This process is regulated by the *Fourier's Law* Eq. 1.1.

$$Q = -\lambda A \frac{d\theta}{dx} \qquad (1.1)$$

where Q represents the *heat flux* in $J \cdot s^{-1}$ (0 Watt), λ is the thermal conductivity of the material ($W \cdot m^{-1} \cdot K^{-1}$), A is the cross-sectional area normal to the direction of heat flow (m²), θ is the temperature (°C or K), and x is the distance measured in

Table 1.2 Typical values for common rocks

	Thermal conductivity $(J \cdot K^{-1} \cdot kg^{-1})$	Volumetric heat capacity $(W \cdot m^{-3} \cdot K^{-1})$
Coal	0.3	1.8
Limestone	1.5–3.0	1.9–2.4
Shale	1.5–3.5	2.3
Wet clay	0.9–2.2	2.4
Basalt	1.3–2.3	2.4–2.6
Diorite	1.7–3.0	2.9–3.3
Sandstone	2.0–6.5	2.0–2.1
Gneiss	2.5–4.5	2.1–2.6
Arkose	2.3–3.7	2.0
Granite	3.0–4.0	1.6–3.1
Quartzite	5.5–7.5	1.9–2.7
Water	0.6	4.18

the direction of flux toward the lowest temperature. An increase in the thermal conductivity allows heat to penetrate deeper into the rock by distributing heat across a larger volume. An increase in the thermal capacity, which is mainly influenced by the density of the considered rock, requires more heat to increase the temperature. Typical values for common rocks are summarized in Table 1.2.

The *thermal conductivity* and the *heat capacity* determine the rock resistance to the variation of temperature, which is a property known as *thermal inertia*:

$$TI = (\lambda * c * \rho)^{\frac{1}{2}} \qquad (1.2)$$

where ρ is the density of the considered material. The most part of the rocks have high values for the heat capacity, while their thermal conductivity is low. This implies that the stored heat is not immediately dissipated and remains stable at depths greater than 10 m, despite the thermal oscillation of temperature on the ground level. The energy is extracted from the soil using *heat exchangers* (Banks 2008), usually installed in a bore or in trench. Every kind of rock has a typical range of values for the *thermal conductivity* and *volumetric heat capacity*. The presence of groundwater table can significantly increase the *thermal conductivity* of every type of rock.

The thermal properties are usually adopted to describe in a complete way a homogeneous material, but the direct measurement of these properties on a rock body is impractical due to its high heterogeneity. This becomes an even greater problem when surveying the properties of rocks over a wide region. An alternative method used to survey the thermal properties of rocks is the study of their emitted electromagnetic spectra. The solar radiation that hits the Earth's surface is partially reflected and absorbed. The reflected component changes with the wavelength of light, and every rock has a specific spectrum. The absorbed component heats the rock and is re-emitted from the surface in the thermal band (8–14 μm). The intensity of emission mainly depends on the surface temperature, which is strongly

Fig. 1.1 Direct versus indirect surveying procedure to detect thermal properties

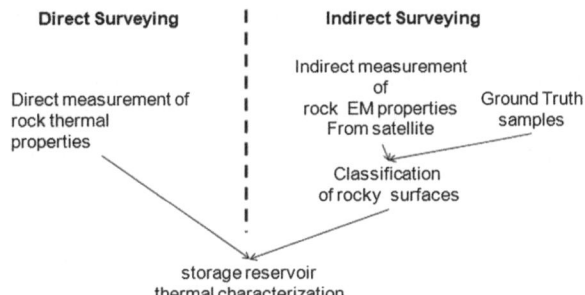

coupled with the average thermal inertia of the rock. The intensities of reflected and emitted radiation are useful features that can be used to thermally characterize the rock. The procedures of direct and indirect surveying of the thermal properties of storage reservoirs are shown in Fig. 1.1.

The *thermal inertia* of rocks can be estimated from satellite images (Drury 2001). This technique is based on the comparison between thermal images taken during the day and night. The temperature of a rock with a low resistance to change in temperature, or low thermal inertia, increases immediately when heated. When the heating stops, the temperature of this rock decreases quickly. The difference (ΔT) between day and night is inversely proportional to the thermal inertia. The apparent thermal inertia (ATI), see Kahle and Alley (1985), also takes the albedo effect into account and is calculated as follows:

$$\text{ATI} = C(1 - a)/\Delta T \tag{1.3}$$

where

a is the albedo;
ΔT is the difference between day and night;
C is a corrective factor that depends on the orientation of the surface and astronomical factors.

Note that the measurements obtained with remote sensing indicate the properties of a shallow layer of rock at a maximum depth of approximately 30 cm, while the deeper rock that constitutes the heat reservoir is hidden from view to the satellite. However, the composition of the surface layer and its other characteristics such as density, porosity, and saturation is strongly correlated with deeper rock and therefore are still able to discriminate the type of substrate. In this way, it is possible to map the lithology of the rocks of the storage heat reservoir using satellite images, even if this information needs to be integrated with ground truth ancillary data. Several factors can influence the measurements made by satellite images. The vegetation and urbanization hide in a non-recoverable way the underground thermal properties. This implies that some portion of the area under study must be masked. Another important factor to take into account is the exposure to sun, which depends on

- the elevation of the sun over the horizon, which is a function of latitude and date/time;
- *Atmospheric conditions*: humidity, air density, altitude;
- *Topography*: altitude and orientation with respect to the sun.

Digital Terrain Models (DTMs) can be used to correct for the effects of these factors.

Nowadays, the application of remote sensed images in the thermal band is widely used in the study of regional resources. In (Nasipuri et al. 2005), the thermal inertia has been successfully used to map the geology. In this case, other methods based on the visual contrast between rocky surfaces failed. In (Nasipuri et al. 2006), thermal images revealed local change of temperature mainly caused by hydrocarbon reservoir. Some recent applications of thermal inertia to the mining field are discussed in Van der Meer et al. (2012). The usefulness of satellite images in the rock/soil classification is well known from the 1970s after the work of (Watson 1975). The satellites for remote sensing are equipped with sophisticated *radiometers* which collect the reflected radiation (albedo) and the emitted radiation of Earth's surface. The radiometers are designed to detect different range of the electromagnetic spectrum. The Moderate Resolution Imaging Spectroradiometer (MODIS[1]) measures the radiations for the visible bands of *Blue* (459–479 nm), *Green* (545–565 nm), *Red* (620–670 nm), the albedo in the near infrared (841–876 nm), and thermal (1,230–1,250 nm). For each one of these bands, the end user can easily obtain a raster image (product) which codes the radiation with spatial resolutions that vary from 250 to 1,000 m. The parameters that characterize a given satellite for remote sensing are as follows:

- *Spatial resolution*: It is defined as the width and number of available bands (from the point of view of electromagnetic spectrum).
- *Temporal resolution*: It is defined as the time (also days) required to acquire a new image of the same area. This resolution plays a key role to solve the change detection problems.
- *Sensibility*: It is defined as the lowest change of the quantity under observation that the instrument is able to detect.

The most used images in the scientific community are produced by commercial and governmental satellite as LANDSAT TM, MODIS, ASTER, and NOAA-AVHRR.

1.4 Potential of Geothermal Energy

The temperature of the rocky ground up to a few meters in depth varies daily. The temperature remains constant at depths greater than 10 m, and this rocky volume constitutes the heat storage reservoir. The thermal inertia of rocks, although

---

[1] http://modis.gsfc.nasa.gov/

fundamental for the "storage" process, is not a quantitative measure of their natural thermal potential. Moreover, the extractable energy is lower than the natural level of energy due to the inefficiency of energy production technologies (Biberacher et al. 2008) and the constraints imposed to maintain sustainable exploitation. For these reasons, a conventional measure of *potential extractable* energy has been introduced that is related to the rock type and to a plant with standard characteristics. The standard plant is a vertical borehole with a depth varying from 80 to 130 m hosting a Borehole Heat Exchanger (BHE) (Johnston et al. 2011). The BHE is formed by a coaxial U pipe that reaches the end of the borehole where a thermo-vector fluid flows, exchanging heat with the soil. The specific Heat Extraction (sHE) (Watt/m) expresses the amount of power that can be extracted for each meter of the pipe and for the standard plant. If the BHE crosses layers with different sHE values, the average value sHE_{agg} can be considered, weighted by the thickness of each layer. The sHE can be directly measured with direct thermophysical measurements on the test site (thermal response test), see Menichetti et al. (2009), or can be estimated by processing remote sensed images by detecting the rock types. The typical values of sHE have been determined for the main rocks considering a standard GSHP plant that works for 2,400 h/year (Verein Deutscher Ingenieure 2001). In (Ondreka et al. 2007), the regional variability of the specific heat extraction has been studied. For a standard plant, the sHE usually varies from 40 to 70 W for each meter/borehole, but special pipes with spiral geometries are able to increase the efficiency (Longhi et al. 2009).

The heat pump is another important component of the geothermal production plant, extracting the heat from the thermo-vector fluid. The underlying principle of the heat pump is that a low boiling point liquid produces steam that releases heat when compressed. The heat is then distributed throughout standard *domestic radiators* to warm the house and its inhabitants. Overall, the plant consisting of the BHE and the heat pump is called a Ground Source Heat Pump (GSHP). A heat pump can also work in the inverse direction, transferring heat from a cooler to a hotter environment against the natural direction of heat flow. Thus, modern heat pumps can provide complete thermoregulation in both the summer and winter periods. The heat pump consumes electric energy, but this electric consumption is lower than the thermal energy extracted by the soil. This aspect justifies the convenience of GSHP. Usually, the efficiency of a heat pump is expressed by the coefficient of performance (COP), defined as the ratio between the heat C generated by the heat pump and the electric energy E that it consumes

$$\text{COP} = \frac{C}{E} \tag{1.4}$$

The COP of GSHPs typically varies from 3 to 4. Following the Eq. 1.4, with COP = 3, it is necessary to supply the heat pump with 2 kW of electric power to obtain 6 kW of thermal power. In other words, 2 kW of electric energy are consumed to transfer 4 kW of renewable energy from the soil to the house. Usually, the COP of a heat pump is not constant and tends to decrease if the temperature difference between the two environments (soil and house) increases. The COP is higher if the heat is transferred toward the cold. As stated in Banks (2008), the GSHPs in domestic plants work under ideal conditions due to the low temperature difference between the soil and the house. The low "delta" results from the high thermal inertia of the soil. A comparison of the operating conditions of the GSHP with the air–water heat pump used in split-type air conditioners can help to understand its operation. During the winter, the GSHP requires less energy to transfer heat from the ground (e.g., at a stable temperature of 10 °C) to the home environment (e.g., with a temperature set point approximately 20 °C) than a standard split requires to transfer heat from the external atmosphere (e.g., an external temperature of −5°). Similarly, during the summer season, it is more efficient to cool a house by transferring energy from the building to the soil, which has a stable temperature (e.g., 10 °C), than it is to operate a split at an external temperature of 30–35 °C. The GSHP configured as a heating system has a COP value of 3.5, while the water–air heat pump has a COP equal to 2.7. In the case of cooling, the GSHP has an energy efficiency ratio (EER) value of 14, while the water–air heat pump has an EER value of 2.4, where the EER is the performance metric for cooling systems.

The GSHP is the most efficient technological solution to the problem of thermoregulation. The cost to install this type of technology can be estimated as 15 k€ (average value), and the payback period varies from 5 to 7 years. An accurate design process is necessary to keep the costs low for both manufacturing and maintenance. In the design process, it is necessary to consider and accurately estimate the energy needs and potential near the installation site. The most expensive component of a GSHP plant is the BHE, which costs 50 Euro for each meter of ground drilling. During the design process, the depth of the BHE, Z_{BHE}', is calculated by dividing the power demand to the BHE, P_{BHE}, by the sHE_{agg} of the installation site:

$$Z'_{BHE} = P_{BHE}/sHE_{agg} + \Delta Z_{GST} \qquad (1.5)$$

where ΔZ_{GST} is a corrective factor obtained by the ground surface temperature (GST) by a regressive function starting from an experimental dataset (Signorelli and Kohl 2004). The ΔZ_{GST} represents the effect of weather conditions. For example, for a P_{BHE} of 5 kW with an sHE of 50 W/m, the BHE should have a depth of 5,000 (W)/50 (W/m) = 100 m before the corrective factor ΔZ_{GST}, which is usually a few meters. For the same production needs, the sHE and climate factors can together lead to variations in the depth of the BHE up to 50 %, with an important effect on costs. For this reason, it is necessary to accurately estimate these factors to minimize the risk of the investment. In (Gemelli et al. 2011), a procedure is presented here that takes environmental factors into account to calculate the depth of the BHE. The inputs required to properly design the GSHP plant are as follows:

- energy demand;
- climate;
- specific heat extraction.

Many systems have been proposed in the scientific literature to exploit the soil as a heat storage reservoir using solar collectors. However, this storage approach has been demonstrated to be inefficient, see the studies of Trillat et al. (2006), Chapuis and Bernier (2008), Eslami-Nejad et al. (2009). A better approach is to design a hybrid plant where the solar collectors are used with a GSHP. The main advantage of this integrated system is a resizing of the geothermal plant, resulting in a reduction in drilling costs and less impact on the soil.

Emissions (mainly caused by processes that transform energy) constitute another important consideration that should be taken into account in every plant. The total emissions of every type of plant are expressed in equivalent kg of CO_2. This indicator is usually adopted to rationally compare technologies/plants. CO_2 emissions are calculated by multiplying the amount of energy consumed by the emission factor (EF), the mass of CO_2 emitted in the production of a given quantity of consumed energy (Killip 2005). The factors for different types of energy are listed in Table 1.3.

The emissions of a GSHP plant are mainly caused by the consumption of electric energy to supply the heat pump. The energy consumption is smaller than that of other plants, representing an advantage for this type of plant. The equivalent kg of CO_2 for a GSHP plant is calculated by multiplying the electric energy consumption by the EF_{EL} of the power grid (Table 1.3). The second part of this book compares different types of plants with GSHPs in several real scenarios. The comparison will take both costs and emissions into account.

1.5 Decision Support Projects

A *decision problem*, in its simplest form, is a question that admits a binary solution: yes or no. The *solution* of the given problem is calculated according to a set of input variables. The complexity of the decision problem depends on the number of decision variables and the number of feasible solutions. The *decision process* can be defined as the analysis that leads to the solution of the decision problem and is conducted with the support of computational tools. The difficulty of a decision problem is expressed by its degree of structure:

- *Non-structured decisions*: These types of decision problems are complex to solve due to the lack of standard procedures for solving the problem.

Table 1.3 Emission factors

Type of energy	Emission factor (kg CO_2/kWh)
Grid electricity	$EF_{EL} = 0.43$
Methane	$EF_{ME} = 0.19$
Coal	$EF_{CO} = 0.30$

- *Structured decisions*: These types of problems are easy to resolve, and the decision maker can follow a well-known optimal procedure to obtain the solution.
- *Semi-structured decisions*: These decisions fall between structured and unstructured decisions. Only a portion of such problems can be addressed using standard and/or well-known procedures.

Most of the decisions concerning geospatial issues are non-structured or semi-structured problems, due to the local/regional variability of geographic variables.

Decision support refers to the tools and information provided by/to people during a decision process, including analytical and graphical representation tools. A *Decision Support System* (DSS) is a computer-based/computer-driven information system that supports decision-making activities. An important criterion for the classification of DSSs is the ability of these systems to output implicit or explicit solutions to a given problem (Sage 1991). Systems that provide implicit solutions are referred to as *passive* DSSs. These types of systems transform data into a form that is intuitive for the decision maker responsible for the final decision. A geographic map is a typical example of passive decision support. In contrast, an *active* DSS provides a decision directly and indicates the action to be taken to accomplish the decision. Recently, DSSs have integrated libraries for data analysis and knowledge extraction to further improve the productivity of decision makers facing *decision processes* (Mladenic et al. 2003).

Decision processes require a work plan, usually referred to as the *decision support project*. The project is necessary to reduce the time and cost of the decision process and to define the approach to the unstructured part of the decision problem. The project places the decision process in a wider operative context (Nyerges and Jankowski 2009). In the geothermal case, the decision support project has the following purposes:

- *Planning*: The process has to lead to a plan to exploit geothermal resources, also considering other renewable and conventional forms of energy.
- *Improvement programming*: This process has to produce a solution to a decision problem, detecting the best solution based on a pros–cons analysis for each available option.
- *Implementation*: It represents decisional support during the initiation of an activity when feedback is available (e.g., periodic evaluation of executed activities).

Frequently, decision-making processes simultaneously have all these goals, corresponding to different levels of development with deliverables at all levels.

The elements of decisional environments are described in the scheme shown in Fig. 1.2. For a given decision problem and its applicative context, a project is defined. This project is formed by a technological environment represented by a DSS, its components, and a set of actions that makes up the decision process leading to the solution of the problem.

Fig. 1.2 Decisional
environment

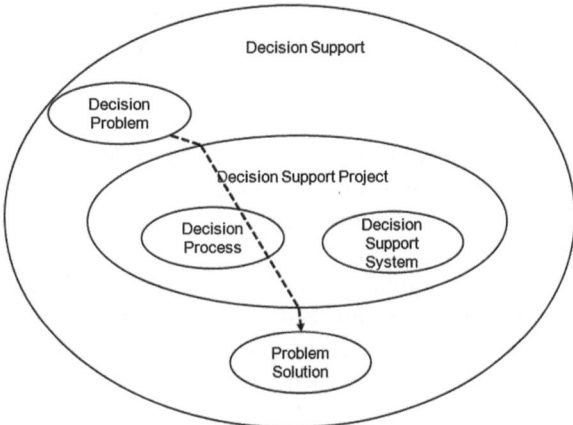

Fig. 1.3 Decision process
workflow

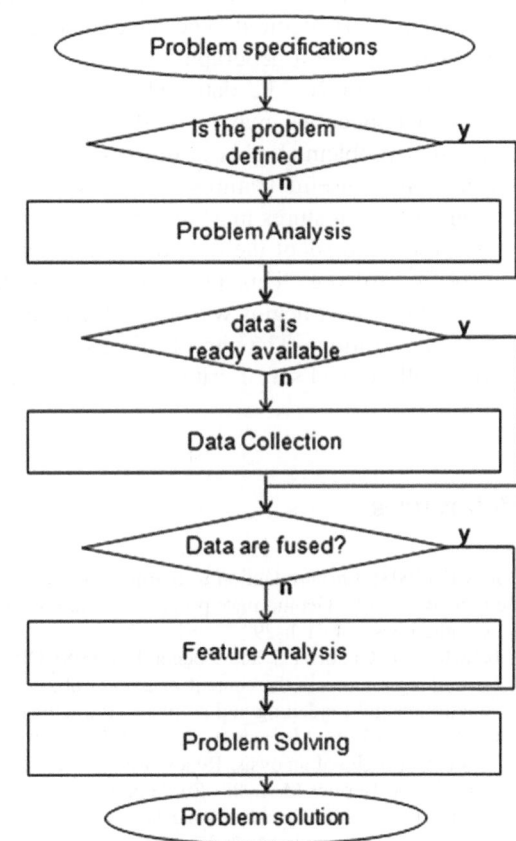

A decision support project includes the development of a *workflow* for the decision process as well as the identification of the DSS technologies used to implement the process. A generic decisional workflow is described below (Fig. 1.3):

- *Project specifications*: As first step, the specifications expressing the purposes and constraints of the "customer" are collected. The purposes and all the measurable features need to be well stated in the specifications without any ambiguity.
- *Problem analysis*: If the purposes and criteria are defined with sufficient clarity, it is possible to proceed with the analysis of the problem by focusing on the expectations of each involved stakeholder. The decision criteria can be identified with variables such as the resource potential, costs, and indicators of specific interest of the stakeholders.
- *Data collection*: If the data are not available or ready to use but rather they must be acquired from several independent external sources, it is necessary to convert them to a common reference structure. This means that a data structure such as a geodatabase is required. The data are organized based on geographic features that characterize a geographic area.
- *Feature analysis*: The data, when they come from heterogeneous and multipurpose collections, may be partly unnecessary or redundant for solving a given decision problem. If this is the case, then it is necessary to proceed with the selection of useful features and eventually to the calculation of new derived features. The features may need to be analyzed at different intermediate steps of the process to adapt the data to current needs.
- *Problem solving*: This phase is separate from the feature analysis, as the problem solving begins with the selected/extracted features and a limited set of feasible solutions. The aim of this task is to find the optimal solution according to a well-defined set of criteria (quantitative and qualitative).

References

Banks D (2008) An introduction to thermogeology. Blackwell, Oxford
Bertani R (2012) Geothermal power generation in the world 2005–2010 update report. Geothermics vol 41:1–29
Biberacher M, Gadocha S and Zocher D (2008) GIS based model to optimize possible self-sustaining regions in the context of a renewable energy supply. In: International congress on environmental modelling and software. Barcelona, Spain
Blum P, Campillo G, Muench W, Koelbel T (2010) CO2 savings of ground source heat pump systems-a regional analysis. Renew Energy 35:122–127
Chapuis S and Bernier M (2008) Étude préliminaire sur le stockage solaire saisonnier par puits géothermiques. In: Canadian solar buildings conference. Fredericton, Canada, pp 14–23
Dickson MH and Fanelli M (2005) Geothermal energy: utilization and technology. Routledge, London
Drury SA (2001) Image interpretation in geology, Routledge, London

Eslami-Nejad P, Langlois A, Chapuis S, Bernier M, and Faraj W (2009) Solar heat injection into boreholes. In: Canadian solar buildings conference. Toronto, Canada

Gemelli A, Mancini A, Longhi S (2011) GIS-based energy-economic model of low temperature geothermal resources: a case study in the Italian Marche region. Renew Energy 36:2474–2483

Haehnlein S, Bayer P, Blum P (2010) International legal status of the use of shallow geothermal energy. Renew Sustain Energy Rev 14:2611–2625

Johnston IW, Narsilio GA, Colls S (2011) Emerging geothermal energy technologies. KSCE J Civil Eng 15(4):643–653

Kahle AB, Alley RE (1985) Calculation of thermal inertia from day-night measurements separated by days or weeks. Photogram Eng Remote Sens 51:73–75

Killip G (2005) Emission factors and the future of fuel, environmental change institute. University of Oxford, Oxford

Longhi S, Cavalletti M, Gemelli A, Kidiamboko S and Mancini A (2009) Low depth geothermal energy: a framework for the design and simulation of U and spiral probes. In: Advanced manufacturing systems for geothermal energy, energy resources. Ancona, Italy

Lund JW, Freeston DH and Boyd TL (2010) Direct utilization of geothermal energy 2010 worldwide review. In: World geothermal congress, international geothermal association. Bali, Indonesia

Menichetti M, Renzulli A, Piscaglia F and Blasi A (2009) Low enthalpy geo-thermal resources: underground temperatures and thermal conductivity, in Advanced manufacturing systems for geothermal energy. Energy Resources, Ancona, Italy

Mladenic D, Lavrac N, Bohanec M and Moyle S (2003) Data mining and decision support: integration and collaboration. Springer, New York

Nasipuri P, Mitra DS, Majumdar TJ (2005) Generation of thermal inertia image over a part of Gujarat a new tool for geological mapping. Int J Appl Earth Obs Geoinf 7:129–139

Nasipuri P, Majumdar TJ, Mitra DS (2006) Study of high-resolution thermal inertia over western India oil fields using ASTER data. Acta Astronaut 58:270–278

Nyerges TL, Jankowski P (2009) Regional and urban GIS: a decision support approach. Guilford Press, New York

Ondreka J, Rusgen MI, Stoberb I, Czurda K (2007) GIS-supported map-ping of shallow geothermal potential of representative areas in south-western Germany—Possibilities and limitations. Renew Energy 32(13):2186–2200

Sage AP (1991) Decision support system engineering, Wiley, NewYork

Signorelli S, Kohl T (2004) Regional ground surface temperature mapping from meteorological data. Glob Planet Change 40(3):267–284

Sudhakar R, Painuly JP (2004) Diffusion of renewable energy technologies: barriers and stakeholders' perspectives. Renew Energy 29:1431–1447

Trillat V, Souyri B, and Achard G (2006) Numerical simulations of ground–coupled heat pumps combined with thermal solar collectors. In: Conference on passive and low energy architecture (PLEA2006). Geneva, Switzerland

Van der Meer FD, Van der Werff HMA, Van Ruitenbeek FJA, Hecker CA, Bakker WH, Noomen MF, Van der Meijde M, Carranza EJM, de Smeth JB, Woldai T (2012) Multi—and hyperspectral geologic remote sensing: a review. Int J Appl Earth Obs Geoinformation 14(1):112–128

Verein Deutscher Ingenieure (2001) Thermische Nutzung des Untergrundes–Blatt 2:erdgekoppelte Warmepumpenanlagen. Beuth Verlag, VDI-Richtlinie 4640

Vine E (2005) An international survey of the energy service company (ESCO) industry. Energy Policy 33(5):691–704

Watson K (1975) Geologic applications of thermal infrared images. In: Proceedings of IEEE, vol 63. Washington D.C., USA, pp 128–137

Chapter 2
GIS-Supported Decision Making

Abstract For the regional scale planning of the renewable energy production, it is required to systematically assess the availability of the resource and the convenience of its exploitation in each site individually. The GIS is the ideal tool to process large amounts of data that subtend the decision problem in a geographic dimension, to synthesize the models, and make them accessible. In this chapter, we present some analytical tools that are included in the architecture of GIS and extend its capabilities as a decision support system. The emphasis is on methods that automate issues such as selection of information, site classification, problems solving. An information system is outlined that has the ability to self-configure to maximize efficacy and efficiency of the data processing.

2.1 System Components

The following state-of-the-art definition by Burrough and McDonnell (1997) is well accepted by the scientific community: "GIS is a powerful set of tools to collect, store, retrieve, process, and represent spatial data in the real world for a defined set of objectives". GIS is also a technological environment characterized by the ability to offer support for decisions, as already recognized by Cowen (1988), who considered GIS as a geographic DSS. GIS is actually also used for *knowledge discovery in spatial data* (a branch of data mining), where it is necessary to find patterns and/or rules hidden in the geographic data, including technologies that strengthen the utility of GIS as DSS. The main components of a geographic DSS are as follows:

- data management subsystem;
- map-based user interface;
- data analysis subsystem;
- decision tools subsystem.

A. Gemelli et al., *GIS to Support Cost-Effective Decisions on Renewable Sources*,
SpringerBriefs in Applied Sciences and Technology,
DOI: 10.1007/978-1-4471-5055-8_2, © The Author(s) 2013

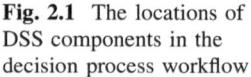
Fig. 2.1 The locations of DSS components in the decision process workflow

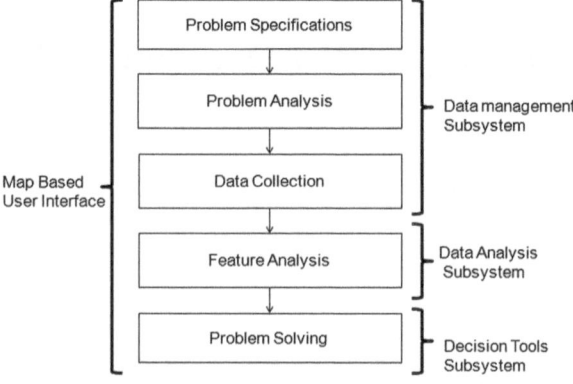

Figure 2.1 shows the relationship of these subsystems with the steps of the decision workflow. Each component will be described in the following parts of this chapter.

2.2 Data Management Subsystem

The *data management* in a GIS environment is based on the logic scheme of a table, or *flat file*. Each record of the table corresponds to a geographic site, and the columns represent the features of each site (Berry 1964). The analysis of a given site is based on a record, while the analysis of a large region composed of many sites requires many records. Studies of single features, which are aimed at detecting variations over a wide region, are conducted on a column. An exploratory study of several features aimed at detecting statistical or mathematical relationships between features is done on all columns or on a selection of columns. In the table, it is possible to include a third dimension that is represented by time. This hypothetical "cube" contains the temporal series for each feature for any given site. The *geographic coordinates* (*latitude, longitude*) are the main features that allow for location-based information management. The data contained in a GIS are mostly georeferred. The georeferred data of a feature are named a GIS layer. This layer can be stored in a dedicated file or in a geodatabase (i.e., PostGIS[1]).

2.3 Map-Based User Interface

The graphical representation of georeferred data is enabled by the concept of layers. A GIS layer is conceptually similar to a geographic *thematic map* that contains all the information related to a specific geographic feature, including the

---

[1] http://www.postgis.org/

geographic location of each data. In computer graphics, a layer is rendered with a *raster image*, which is a rectangular grid of pixels where each pixel represents a limited geographic area. The value of a pixel represents the value of a feature such as the average altitude in the pixel area. The domains of these values can be discrete, real, string, or null in cases in which the value is not available. Standard symbols and/or colors are adopted to represent the value of each pixel to immediately render the map content to the user. For example, a raster image is produced by satellite acquisition or can be derived by the processing of other raster features. Some feature can be described by geometric primitives as points or polygons. An example of this type of feature is the country/province boundaries related to the total resident populations or average salaries. This type of representation has a higher efficiency from the point of view of storage and rendering but introduces an approximation in which the polygon is implicitly internally homogeneous and in which the changes only occur on its perimeter.

2.4 Data Analysis Subsystem

The *Data Analysis Subsystem* provides support to the *feature analysis* phase of the decision workflow. This subsystem consists of methods based on *vectorial analysis* where each row of the flat file is considered as a vector of n features. The vectors x can be projected as points in a multi-dimensional space X^n, defined as the space of the features. In X^n, the positions of the points highlight groups, hyperplanes of separation between groups, directions of maximum variability and other notable elements which are the subject of study of feature analysis. This paragraph presents the methods of the data analysis subsystem.

The data analysis subsystem includes two main categories of methods:

- methods for feature selection;
- methods for the extraction of predictive models.

The aim of Feature Selection (FS) (Dougherty 2005; Singhi and Liu 2006) is to select the most important features to build the predictive model for a target feature, excluding the useless features. This has the consequence to reduce the costs of data acquisition and the computational resources needed. If a lower limit is set for the required accuracy of the predictive model, a FS algorithm proceeds by creating several subsets of features and then selects the subset that reaches the required accuracy with the smallest amount of features. A family of FS algorithms follows the *heuristic approach* in searching the optimal subset, which is a simple approach but do not provide any guarantee of optimality, and moreover, become progressively computationally infeasible as the number of features grows (Guyon and Elisseeff 2003). Another family of FS algorithms follows the *deterministic approach* in which the optimal subset of features is precisely determined without room for random variation. This property is based on the ability to weight each feature according to its relevance. The optimal subset of features is constructed by sequentially adding features starting from the most important (the feature with the highest weight) until the required accuracy is reached. The Feature Extraction (FE) algorithms belong to the *deterministic approach*. The FE aims to build new and more expressive features which are linear combinations of original ones. The FE looks for (linear or nonlinear) mappings of the original features into features more relevant for the prediction, performing a space transformation. If the mapping is linear, then the FE problem can be viewed as finding the mapping matrix A composed of $n' < n$ orthonormal vectors $\varphi_1, \ldots \varphi_{n'}$, transforming the feature space X^n in the reduced space $X^{n'}$, such that $\forall x \in X^n, \exists x' = \sum_{i=1}^{n'} x \cdot \varphi_i$. Eigenvalues λ of the orthogonal transformation are treated as relevance weights of the new features represented by the eigenvectors of A.

The extracted features are oriented in directions parallel or normal the direction that best separate the classes.. In Fig. 2.2, the FE methodology is applied to a two-class classification problem. x_1, x_2 and y_1, y_2 are original and extracted new features, respectively. y_1 is a redundant feature ($\lambda_1 = 0$), giving no contribution to the classification, while y_2 is an informative one ($\lambda_2 > 1$). FE methods differentiate each other for the algorithm used to obtain the mapping matrix, a survey of FE techniques is in Diamantini and Potena (2007). A family of FE algorithms for classification goal is the one based on the Decision Border Feature Extraction (DBFE) method (Lee and Landgrebe 1993), in which the directions normal to the decision border are extracted and used to build the mapping matrix called Effective Decision Border Feature Matrix (EDBFM). The Bayesian Vector Quantizer (BVQ) Feature Extraction (BVQFE) (Diamantini and Potena 2007) applies the approach of minimization of the Bayesian misclassification risk to border detection and feature extraction.

Another FE method is the Linear Discriminant Analysis (LDA). This algorithm resolves a maximization problem whose objective function is commonly expressed by the *Generalized Eigenvalue Problem*: $S_b A = S_w A \Lambda$. Where Λ is a diagonal matrix whose entries are the eigenvalues, S_b and S_w, which are metric matrices that represent measures to be maximized and to be minimized, respectively. In the classical LDA, S_b is the between classes scatter matrix, and S_w is the within classes scatter matrix. The objective is to maximize $S_w^{-1} S_b$ product by an opportune choose of A, as expressed by: $A = \arg\max_A \{ \mathrm{trace}(S_w^{-1} S_b) \}$. The Orthogonal Linear Discriminant

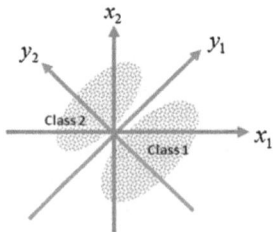

Fig. 2.2 Feature extraction

Analysis (OLDA) algorithm, by Ye (2005), is a variant of LDA. In OLDA, the mapping matrix A is computed through the simultaneous diagonalization of scatter matrices. It results that OLDA is applicable even when scatter matrices are singular.

A problem of all FE techniques is that they weight a new set of features, which are linear combinations of original ones and are not human-readable. The Goal Oriented Feature Ranking (GOFR) has been proposed in Gemelli et al. (2009b), as a novel technique based on FE algorithms integrated with a novel procedure that retrieves the individual weight of the original features a further manipulation of by mapping matrix A: firstly, the eigenvectors are weighed by multiplying them by the respective eigenvalues, and then the corresponding components of weighed eigenvectors are summed (in the absolute values). Resulting values are the individual contributions (or weights) of each original feature into the new set of features. The weight allows to sort the original feature by their discriminative power (Fig. 2.3) relatively to the classification goal.

In the following, we report the pseudo-code of the GOFR algorithm, see also Fig. 2.3:

Let $X = \{x_1, x_2, \ldots, x_n\}$ be the m-dimensional original feature space, and let G the given goal (the feature to be predicted).

1. Apply the FE algorithm on <X,G>, and let A be the output matrix;
2. Let $Y = \{y_1, y_2, \ldots, y_n\}$ be the eigenvectors of A, and $\Lambda = \{\lambda_1, \lambda_2, \ldots, \lambda_n\}$ be the related eigenvalues.
3. For $j = 1$ to n, Compute the weight $w_j = c_i \sum_{i=1}^{n} |\lambda_i||y_{ij}|$

In the formulas, the factor c_i is optional, representing the individual cost for each feature, incorporating acquisition cost, sensor price, energy consumption, and

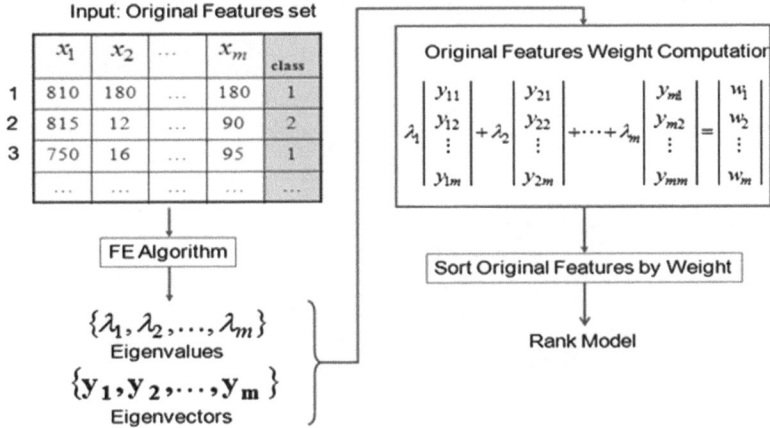

Fig. 2.3 Goal-oriented feature ranking algorithm

so on. The costs transform the rank model in a cost function. The output of this GOFR is the rank model (RM), defined as the following tuple:

$$RM = \{\text{dataset}, \text{goal}, \text{rank}\}$$

where the *goal* is the target feature chosen as class, and *rank* is a vector of pairs *<feature, weight>*.

Every FE algorithm can be applied as core of GOFR. In Diamantini et al. (2010), the applicability of EDBFM as FE algorithm at the core of GOFR is demonstrated. A complete benchmarking of BVQFE, LDA, and OLDA algorithms is proposed in Gemelli (2010). The effectiveness of the GOFR in the GIS environment has been successfully tested in Gemelli et al. (2009a) and in Gemelli (2010), with positive fallback on the running costs of informative system.

The *predictive algorithms* such as the classifiers also are tools of the data analysis subsystem. The aim of classification is to derive a rule or set of rules to correctly predict a class (*target feature*) for each vector of the dataset. The rule should minimize the risk of misclassification (Friedman 1997). A common application of classification to GIS is the ability to predict a hidden or not directly measurable geographic feature on the basis of a set of known features (Fig. 2.4). The prerequisite is that a cause–effect relationship exists. A classical GIS application is the classification of remote sensed images, for example, predicting the land use, based on spectral features acquired by multi-spectral sensors. Classification algorithms are commonly grouped into two categories: *supervised* and *unsupervised* approaches. The first approach requires a training set to build the set of rules that will be used to predict the class.

During last years, the researchers developed a wide variety of supervised classifiers (Witten and Frank 2005): the main algorithms are the support vector machine (SVM) (Vapnik 2005), *maximum likelihood* (Duda et al. 2000), *AdaBoost* (Sutton 1995), *neural networks* (Pasika et al. 2009). *Adaptive Boosting*, often known as *AdaBoost* is one of the most used boosting algorithm to enhance the performance of a generic classifier by creating strong hypothesis from weak

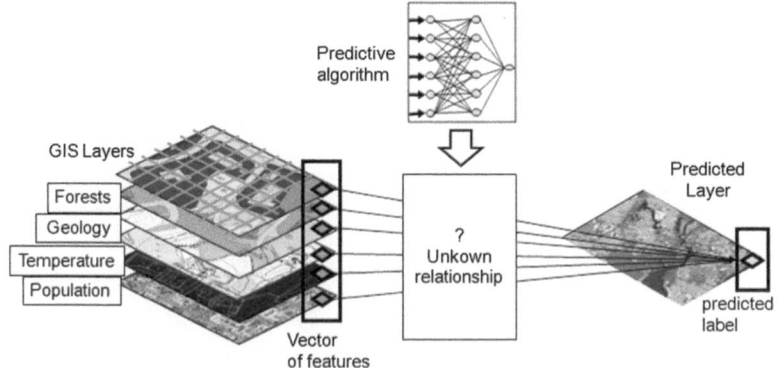

Fig. 2.4 Geographic feature prediction task

classifiers' combinations (e.g., LMS, SVM, Perceptron, Decision Tree). The key concept of AdaBoost is to iteratively focus on difficult patterns by increasing the weights of misclassified training patterns and decreasing the weights of correctly classified training patterns. The first problem was explored by Freund and Schapire (1997; Schapire and Singer (1999) using different boosting variants such as Discrete AdaBoost, Real AdaBoost (used in this work), and Gentle AdaBoost (Friedman et al. 2000) which differ in the learning schemes but not in the computational loads. The Real AdaBoost algorithm works as follows: each labeled training pattern x receives a weight that determines its probability of being selected for a training set for an individual component classifier. Starting from an initial (usually uniform) distribution Dt of these weights, the algorithm repeatedly selects the weak classifier $h_t(x)$ that returns the minimum error according to a given error function. If a training pattern is accurately classified, then its chance of being used again in a subsequent component classifier is reduced; conversely, if the pattern is not accurately classified, then its chance of being used again is increased. In this way, the algorithm takes the approach of modifying the distribution Dt by increasing the weights of the most difficult training examples in each iteration. The selected weak classifier is expected to have a small classification error on the training data. The final strong classifier H is a weighted majority vote of the best T (number of iterations) weak classifiers $h_t(x)$:

$$H_{(x)} = \text{sign}\left(\sum_{t=1}^{T} \alpha_t h_t(x)\right) \tag{2.1}$$

It is important to notice that the complexity of the strong classifier depends only on the weak classifiers. The AdaBoost algorithm has been designed for binary classification problems. To deal with non-binary results, there are two common options:

- one against all (OAA);
- one against one (OAO).

The OAA allows to deal AdaBoost with multi-classes problem; given C classes, the algorithm creates C different binary classifiers. At class ith, the +1 label is assigned and to all remaining classes −1. Then, the Eq. 2.2 is modified as follows:

$$H(x, i) = \text{sign}\left(\sum_{t=1}^{T} \alpha_t h_t(x)\right) \tag{2.2}$$

with $i = 1, ..., C$. For each instance x, a voting vector $V(x)$ is built:

$$V(x) = [H(x, 1) \ldots H(x, C)].$$

The following operator is applied to extract the winner class:

$$\max(V(x)) = L$$
$$L = \{i \in N, \quad i = 1, \ldots, C\}$$

The OAO algorithm differs from the OAA by the set of classifiers: given C classes, the algorithm creates $C \cdot (C-1)/2$ different binary classifiers; this number represents the combination without repetitions. The Eq. 2.3 is modified as follows:

$$H(x, i, j) = \text{sign}\left(\sum_{t=1}^{T} \alpha_t h_t(x) \right) \tag{2.3}$$

with $i, j = 1, \ldots, C(C-1)/2$ and $i \neq j$. For each instance x, a voting matrix $V(x)$ is built:

$$v(x) = \begin{bmatrix} 0 & H(x,1,2) & H(x,1,3) & \cdots & \cdots & H(x,1,C) \\ \vdots & 0 & H(x,2,3) & \cdots & \cdots & H(x,2,C) \\ \vdots & 0 & 0 & H(x,3,4) & \cdots & H(x,3,C) \\ \vdots & \vdots & \vdots & \ddots & \vdots & \vdots \\ \vdots & \vdots & \vdots & 0 & 0 & H(x,C-1,C) \\ 0 & \cdots & \cdots & \cdots & \cdots & 0 \end{bmatrix}. \tag{2.4}$$

$$V_{\text{sum}}(x) = \left[\sum_{i=1}^{C}\sum_{j=1}^{C} f(V_{ij}(x), 1) \quad \cdots \quad \sum_{i=1}^{C}\sum_{j=1}^{C} f(V_{ij}(x), C) \right]^{T}$$

$$f(V_{ij}(x), k) = \begin{cases} V_{ij}(x) & \text{if}(i == k \ \vee \ j == k) \\ 0 & \text{else} \end{cases}$$

V is a C C matrix and Vsum is C·1 vector; the following operator is applied to extract the winner class:

$$\max(V_{\text{sum}}(x)) = L$$
$$L = \{i \in N, i = 1, \ldots, C\}.$$

It is evident that the OAO scheme is suitable when the class number is low because the number of classifiers quickly grows up. The OAO can be also integrated with a weight represented by the spectral separability; this can be accomplished by modifying the definition of V matrix:

$$
\begin{bmatrix}
0 & w_{1,2}H(x,1,2) & w_{1,3}H(x,1,3) & \cdots & \cdots & w_{1,C}H(x,1,C) \\
\vdots & 0 & w_{2,3}H(x,2,3) & \cdots & \cdots & w_{2,C}H(x,2,C) \\
\vdots & 0 & 0 & w_{3,4}H(x,3,4) & \cdots & w_{3,C}H(x,3,C) \\
\vdots & \vdots & \vdots & \ddots & \vdots & \vdots \\
\vdots & \vdots & \vdots & 0 & 0 & w_{C-1,C}H(x,C-1,C) \\
0 & \cdots & \cdots & \cdots & \cdots & 0
\end{bmatrix}
$$

$$(2.5)$$

where w_{ij} is the spectral separability between the i and j classes. In the framework, an instance x can be also labeled as unclassified; this happens if all the weights $H(x)$ or $H(x, i, j)$ are below the zero.

Traditionally used pixel-based classification methods are based on conventional techniques. Although this approach performs well, it has a limited ability to resolve inter-class confusion. The pixel-based approach is strongly limited by the type of output; the final product of the pixel-based approach is usually represented by a raster image that does not allow for the extraction of meaningful information due a lack of topological information. However, this approach tries to exploit the benefits of supervised classifiers such as AdaBoost. Starting from a set of spectral band layers, it is possible to obtain an image where each pixel belongs to a class. Pixel-based classification has a drawback in the presence of salt-and-pepper noise and the impossibility of extracting complex classes from heterogeneous areas. In the elaboration, we applied a majority filter to reduce the presence of salt-and-pepper noise as much as possible.

The *accuracy of classification* is a performance indicator that is critical in the data analysis subsystem. The accuracy is usually calculated as the number of correct predictions over the total number of available known points. The target of 100 % is usually a "chimera". The accuracy is influenced by the quality of the data (e.g., the types of classes and their spectral separability) and the efficacy of the classifier. In some cases, reducing the set of features can increase the accuracy of classification. This effect is well known and is defined as the *curse of dimensionality* (Hastie et al. 2001). The GOFR (Gemelli et al. 2009a), under the assumption that the goal coincides with the classification, produces an optimal reduction in the number of features and minimizes the curse of dimensionality. The algorithms of features extraction, feature ranking, feature selection, and the *predictive algorithms* belong to the disciplinary field named *Data Mining* (Witten and Frank 2005) defined as "the extraction of implicit, previously unknown, and potentially useful information from data". The extraction of information is considered as *new knowledge* that is automatically extracted by an appropriate algorithm; for this reason, data mining is also known as *Automatic Learning*. Today, data mining is widely adopted in a variety of fields (Miller and Han 2009) (Gaber 2010). There is evidence that automatic learning has a greater power to discover knowledge in the analysis of geographic data than any other approach because it is

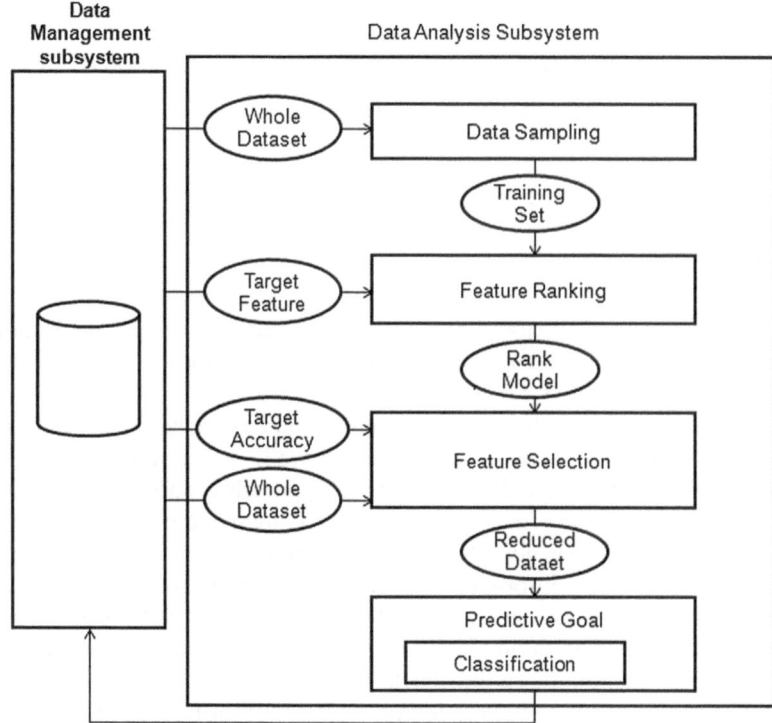

Fig. 2.5 Data analysis subsystem

the only approach able to investigate the complexity of geographic data. Over the years, the potential of the automatic approach to study geographic phenomena has been studied and has given rise to a field of research called *Geocomputation* (Openshaw 1998) a (Fischer and Yee Leung 2010).

In a GIS environment, the presence of a data analysis subsystem is structural (Gemelli et al. 2009a). This subsystem is composed of a set of modules as shown in Fig. 2.5, with each module formed by a set of libraries for (a) data sampling, (b) feature ranking, (c) feature selection, and (d) predictive data mining tasks. This subsystem also plays a role in the decision-making process to optimize the internal processes of GIS by reducing the computational load (time to output and memory occupancy).

2.5 Decision Tools Subsystem

A distinction exists between data analysis tools and decisions tools. The decision tools focus on synthesizing features and provide directly a solution to the decisional problem, also embedding the preferences of the human operator. The advantage provided by the *Decision Tools Subsystem* is to automate the task assigned to a human operator when the decision problem to solve is too complex in terms of the number of features involved, or it must be resolved repeatedly in a large number of cases and there are real-time requirements. In these difficult conditions, the *Decision Tools Subsystem* supports the human operator to make objectives choices.

The decision process is supported by the Multi-Criteria Decision Analysis (MCDA) by Zionts (1979), see also (Wang et al. 2009a). MCDA solves the problem based on a linear weighted combination of features considered as decisional criteria.

Given a set of alternative solutions $i = 1,..., m$ and a set of decisional criteria $j = 1,..., n$, for each alternative solution i, the score s_i is calculated as:

$$s_i = \sum_{j=1}^{n} w_{ij} \cdot c_{ij} \quad i = 1,...,m \tag{2.6}$$

where c_{ij} is the normalized value of the criterion. w_{ij} is the weight of criterion.

The alternative that is assigned the highest score represents the solution to the problem. The decision maker plays a key role in that the subjective preferences of the decision maker implemented in the criteria weights have a determining impact on the score, with the result that decisional problems do not have unique solutions.

The MCDA is widely used in GIS environments. Here, the criteria are a set (or subset) of features. Usually, a GIS layer is associated with each criterion. Given the set of weights, the Eq. 2.6 is applied and the output is usually a new layer that represents the final solution. Different methodologies have been proposed based on the MCDA. An interesting survey of MCDA variants has been conducted (Nyerges and Jankowski 2009). The weight is usually considered static, but in one case (Malczewski 2011) the weight is defined as function of the geographic area. The popularity of MCDA methods in GIS is due to their ease of use, as they essentially consist of simple algebraic operations among layers.

In a MCDA application, there are the following main components:

- the decision to reach;
- the human decision maker (who set the weights);
- the set of alternative decision;
- the set of criteria;
- the set of results/consequence related to each alternative.

These procedures can also be iterative when the decision maker decides to change weights after evaluating the obtained results. The solution is the best compromise for the decision maker considering the assigned weights.

Energy planning is usually performed without considering spatial relationships, and all sites are considered to represent in a unique context. This approach worked well until several years ago, as the energy market has been totally dominated by the centralized approach to production based on fossil fuels and hydroelectricity. However, the recent exponential growth in renewable energies is now imposing

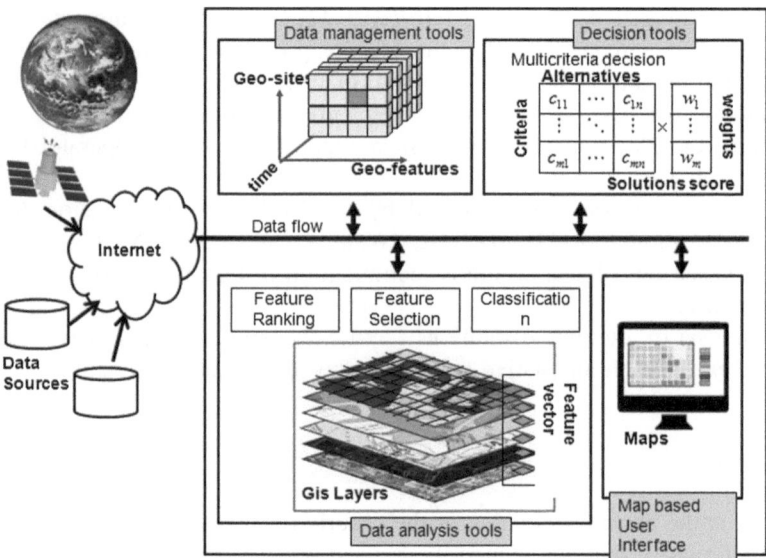

Fig. 2.6 GIS-based decision support system

new rules. The new paradigm is "many small plants for many end users that tend to locally drain locally produced energy". For this reason, energy planning for renewable energies should be based on specific information tailored to characterize the unique context of each area. The decisional environment of renewable energies is complex due to the presence of a large set of features that vary over space and time. The complexity increases if the economic aspects of renewable energies are also considered, including the (a) high costs of plants and (b) the reduced return on investment. For this reason, it is necessary to conduct an accurate analysis to evaluate the convenience of investing in renewable energy at a particular geographic site. GIS is a perfect tool for energy planning at the national/regional scale. In Voivontas et al. (1998) and Ramachandra and Shruthi (2007), the GIS has been used to optimize the planning of renewable energies. There are many applications of this type in developing countries: GIS has been used to calculate monetary and emission savings (Yue and Yang 2007) and to plan the energy supply in a rural area that is usually considered off-grid (Byrne et al. 2007). In the GIS environment, the MCDA allows for rapid evaluation of a large set of conditions for each area and produces the best solution based on socioeconomic, geographic, and physical information. The MCDA has been successfully applied by Gret-Regamey and Hayek (2012) to energy planning with a particular focus on the identification of production sites. In Omitaomu et al.'s (2012) study, they divided the territory into a 100×100 m grid and evaluated the acceptability to host a renewable energy production plant.

The combination of GIS equipped with DSS components and remote sensing (Fig. 2.6) is a successful strategy to solve energy planning problems. Remote sensing is interesting due to the availability of a high-resolution dataset (temporal and spatial resolution). Remote sensing allows for the immediate measurement of some physical variables of large geographic areas that are strongly coupled with renewable energy (Wang et al. 2009b, Geib et al. 2012). Remote sensing technology has continuously progressed, and the accuracy and precision of the resulting data are becoming similar to those of direct/on-site surveys.

References

Berry BJL (1964) Approaches to regional analysis: a synthesis. Ann Assoc Am Geogr 54:2–11

Burrough PA, McDonnell RA (1997) Principles of geographical information systems. Oxford University Press, Oxford

Byrne J, Zhou A, Shen B, Hughes K (2007) Evaluating the potential of small-scale renewable energy options to meet rural livelihoods needs: a GIS- and lifecycle cost-based assessment of Western China's options. Energy Policy 35:4391–4401

Cowen D (1988) GIS versus CAD versus DBMS: what are the differences? Photogram Eng Remote Sens 54(11):1551–1555

Diamantini C, Potena D (2007) A study of feature extraction techniques based on decision border estimate. In: Liu H, Motoda H (eds) Computational methods of feature selection. Chapman & Hall, London

Diamantini C, Gemelli A, Potena D (2010) Feature ranking based on decision border. In: International conference of pattern recognition (ICPR), IEEE

Dougherty ER (2005) Feature-selection overfitting with small-sample classifier design. IEEE Intell Syst 20(6):64–66

Duda RO, Hart PE, Stork DG (2000) Pattern classification. Wiley, New York

Fischer MM, Leung Y (2010) Geocomputational modelling: techniques and applications. Springer, Heidelberg

Freund Y, Schapire RE (1997) A decision-theoretic generalization of on-line learning and an application to boosting. J Comput Syst Sci 1(55):119–139

Friedman J (1997) On bias, variance, 0/1—loss, and the curse-of-dimensionality. Data Min Knowl Disc 1:55–77

Friedman J, Hastie T, Tibshirani R (2000) Additive logistic regression: a statistical view of boosting. Ann Stat 38(2):337–374

Gaber MM (2010) Scientific data mining and knowledge discovery. Springer, Berlin

Geib C, Taubenbock H, Wurm M, Esch T, Nast M, Schillings C, Blaschke T (2012) Remote sensing-based characterization of settlement structures for assessing local potential of district heat. Remote Sens 3:1447–1471

Gemelli A (2010) Goal oriented feature ranking for intelligent geospatial decision support systems. Ph.D., Università Politecnica delle Marche

Gemelli A, Diamantini C, Potena D (2009a) A feature ranking component for GIS architecture. In: D'Atri A, Saccà D (eds) Information systems: people, organizations, institutions, and technologies. Physica, Wurzburg

Gemelli A, Diamantini C, Potena D (2009b) A novel feature ranking modeling in GIS context: addressing complexity and cost issues. In: International conference on geoinformatics

Gret-Regamey A, Hayek UW (2012) Multicriteria decision analysis for the planning and design of sustainable energy landscapes. In: Stremke S, Van Den Dobbelsteen A (eds) Sustainable energy landscapes. CRC Press, Boca Raton, pp 111–132

Guyon I, Elisseeff A (2003) An introduction to variable and feature selection. J Mach Learn Res 3:1157–1182

Hastie T, Tibshirani R, Friedman J (2001) The elements of statistical learning: data mining, inference, and prediction. Springer, New York

Lee C, Landgrebe DA (1993) Feature extraction based on decision boundaries. IEEE Trans Pattern Anal Mach Intell 15(4):388–400

Malczewski J (2011) Local weighted linear combination. Trans GIS 15(4):439–455

Miller HJ, Han J (2009) Geographic data mining and knowledge discovery. CRC Press, Boca Raton

Nyerges TL, Jankowski P (2009) Regional and urban GIS: a decision sup-port approach. Guilford Press, New York

Omitaomu OA, Blevins BR, Jochem WC, Mays GT, Belles R, Hadley SW (2012) Adapting a GIS-based multicriteria decision analysis approach for evaluating new power generating sites. Appl Energy 96:292–301

Openshaw S (1998) Building automated geographical analysis and explanation machines. In: Longley PA et al (eds) Geocomputation: a primer. Wiley, Chichester

Pasika H, Haykin S, Clothiaux E, Stewart R (2009) Neural networks for sensor fusion in remote sensing. In: International joint conference on neural networks, IEEE, pp 2772–2776

Ramachandra TV, Shruthi BV (2007) Spatial mapping of renewable energy potential. Renew Sustain Energy Rev 11(7):1460–1480

Schapire RE, Singer Y (1999) Improved boosting algorithms using confidence rated predictions. Mach Learn 3(37):297–336

Singhi KS, Liu H (2006) Feature subset selection bias for classification learning. In: International conference on machine learning, pp 849–856

Sutton CD (1995) Handbook of statistics. Elsevier, Amsterdam

Vapnik VN (2005) The nature of statistical learning theory. Springer, New York

Voivontas D, Assimacopoulos D, Mourelatos A, Corominas J (1998) Evaluation of renewable energy potential using a GIS decision support system. Renew Energy 13(3):333–344

Wang JJ, Jing YY, Zhang CF, Zhao JH (2009a) Review on multi-criteria decision analysis aid in sustainable energy decision-making. Renew Sustain Energy Rev 13:2263–2278

Wang H, Leduc S, Wang S, Obersteiner M, Schill C, Koch B (2009b) A new thinking for renewable energy model- Remote sensing-based renewable energy model. Int J Energy Res 33:778–786

Witten IH, Frank E (2005) Data mining: practical machine learning tools and techniques. Elsevier, Amsterdam

Ye J (2005) Characterization of a family of algorithms for generalized discriminant analysis on undersampled problems. J Mach Learn Res 6:483–502

Yue CD, Yang GGL (2007) Decision support system for exploiting local renewable energy sources: a case study of the Chigu area of southwestern Taiwan. Energy Policy 35:383–394

Zionts S (1979) MCDM: if not a roman numeral, then what? Interfaces 9(4):94–101

Part II
Case Study

Chapter 3
GIS-Supported Decision Making for Low-Temperature Geothermal Energy in Central Italy

Abstract The project specifications of a geospatial decision support system are described in this chapter. The system is dedicated to the low-temperature geo-thermal energy, and it allows to build spatial models of the resource and to evaluate the cost-benefit of its exploitation in a wide region of central Italy. The architecture of the system and the computational workflow are depicted. The design is based on a thorough study of data analysis tools.

3.1 Project Specifications

The second part of this book proposes a case study of LTGE exploitation over a wide region. The natural potential and economic features of this resource are systematically calculated at each site and then mapped. Afterward, some of these features are used as decision criteria to decide the optimal plant for domestic thermoregulation in each site.

The region under study is located in Central Italy and extends approximately 26,000 km^2, covering the whole territory of *Marche* and *Umbria* and partially covering the territories of *Tuscany, Emilia Romagna, Lazio*, and *Abruzzi*. LTGE is hypothetically able to fulfill the energy needs for domestic thermoregulation in this area, but the suitability of this resource must be determined site by site. The climate varies significantly across this region, extending from the Adriatic Sea, where the climate is warm, to the Apennine Mountains, with a continental climate. The maximum difference in elevation is 2,800 m. Figure 3.1 shows that the study region projected in the UTM system zone 33 and datum WGS84 is shown with overlaid administrative boundaries.[1] The region is bounded between the following kilometric coordinates: UPPER LEFT X = 244918.022, UPPER LEFT Y = 4876495.896, LOWER RIGHT X = 411818.022, LOWER RIGHT Y = 4694795.896.

---

[1] www.istat.it

A. Gemelli et al., *GIS to Support Cost-Effective Decisions on Renewable Sources,*
SpringerBriefs in Applied Sciences and Technology,
DOI: 10.1007/978-1-4471-5055-8_3, © The Author(s) 2013

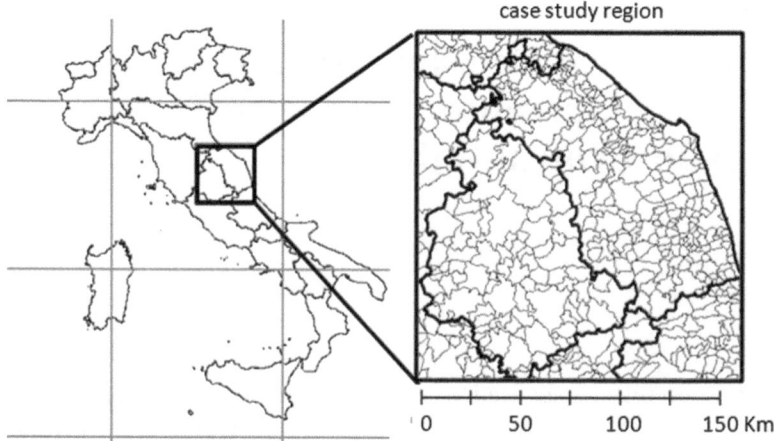

Fig. 3.1 Study region

The results of this study will be represented by a set of numerical spatial models:

- a low-temperature geothermal energy potential model of the region;
- cost-benefit indicators of geothermal plants for domestic thermoregulation and comparison with alternative plants;
- decision models aimed at supporting different categories of stakeholders to decide whether to invest in geothermal resources.

These models will be produced in a GIS environment with the following requirements:

- the models are calculated at each site in the region, and the results are represented on maps;
- the cost-benefit assessments expected for each type of energy production plant shall consider the efficiency of the production technologies, the local potential extractable geothermal resources, and the energy needs;
- the decision support system shall be the implementation of the schema illustrated in Fig. 2.6 and realizing the processing workflow of Fig. 1.3;
- the process shall be easily applied to other regions and to other sources of energy once the decision criteria have been defined.

3.2 Decision Process

The decision-making process is adapted to the geothermal problem according to the workflow of Fig. 3.2.

The workflow of Fig. 1.3 adapted to geothermal resource consists of the steps summarized in Table 3.1.

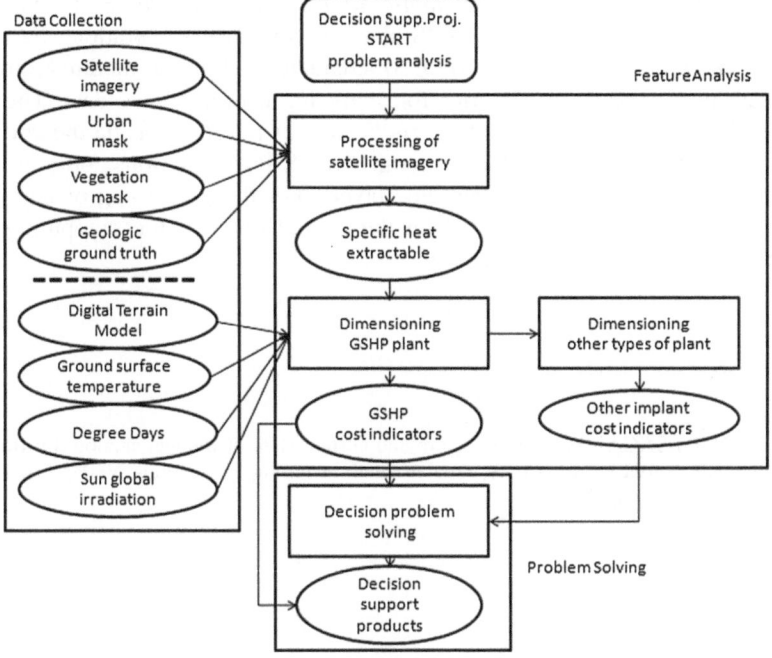

Fig. 3.2 Decision process workflow of geothermal decision process

Table 3.1 Workflow steps

Data collection

The data are collected from multiple sources and then transferred to a common data structure.
  The data serve two main purposes:
  the satellite images are to build the geothermal model
  the data on energy needs are to size the thermoregulation plants

Feature analysis

Three subphases are
  processing of satellite images to obtain the model of the geothermal potential
  comparing analysis of energy needs and geothermal potential for sizing the GSHP system and
    calculating cost indicators
  sizing alternative plants for thermoregulation, equivalent to the GSHP for the amount of
    energy delivered, but based on different energy sources

Problem solving

Multicriteria decision analysis application: using chosen features as decision-making criteria and
  representing results in a map format
grouping the maps in a decision support atlas. Dividing the products into two groups:
  o maps that provide passive support to the decision making
  o maps that provide active support to the decision making

In this work, the *System for Automatic Analysis Geoscientific* (SAGA) (Conrad 2006) software has been used as a GIS platform for process operations. SAGA[2] is a *Free Open Source Software* (FOSS) characterized by a modular structure enhanced by a large set of libraries for analyzing georeferenced data. The functions of SAGA are accessed via the *graphical user interface* or via the command line, even with batch script files. Generally, the SAGA functions apply algebraic operators to the layers to calculate new layers. For the processing of non-georeferenced data or data abstracted from the spatial context, the computing environment of MATLAB[3] has been used.

Reference

Conrad O (2006) SAGA—program structure and current state of implementation. In: Bohner J, McCloy KR, Strobl J (eds) SAGA—analysis and modelling applications. Verlag Erich Goltze GmbH, Germany, pp 39–52

---

[2] www.saga-gis.org
[3] MathWorks, Matlab, 2012, software.

Chapter 4
Data Collection

Abstract The input data to the decision support systems are acquired from multiple sources and transferred to a common data structure. The data are selected on the basis of known relationship between the geothermal potential and the energy demand, which shall, later in the book, be calculated from these data. The choice of data sources is based on criteria of quality and accessibility; the emphasis is on remote sensing systems since they have a wide geographic coverage and are frequently updated.

The data have been collected from multiple sources and can be distinguished into the following categories (Table 4.1).

4.1 Remote Sensing Data Resources

The remote sensing data, originated from *Moderate Resolution Imaging Spectroradiometer* (MODIS), are used to map the lithology and the geothermal potential in detail. MODIS is a payload scientific instrument launched into Earth orbit by NASA in 1999 on board of the Terra (EOS AM) satellite and, in 2002, on board the Aqua (EOS PM) satellite. The MODIS measures the emissivity of the Earth's surface in different bands of the electromagnetic spectrum. MODIS has a viewing swath width of 2,330 km and views the entire surface of the Earth every one to two days. Its detectors measure 36 spectral bands, and it acquires data at three spatial resolutions: 250, 500, and 1,000 m. The 1,000 m resolution products have been considered for this study. Among the several products derived from MODIS and

Table 4.1 Main data categories

Data category	Main purpose for acquisition
Remote sensing data	Geothermal potential modeling
Vegetation coverage and land use	Geothermal potential modeling
Data of energy demand	Sizing the GSHP thermoregulation plant
Digital terrain model	Sizing the GSHP thermoregulation plant
Administrative boundaries	Sizing the GSHP thermoregulation plant

Table 4.2 MODIS acquired products

MOD11A2

The MODIS *global land surface temperature* (LST) data are composed from the daily 1-km LST product and stored on a 1 km sinusoidal grid as the average values of clear-sky LST during an 8-day period. Both the LST night and LST day layers have been acquired for this project

MOD13A3

The MODIS global vegetation indices are designed to provide consistent spatial and temporal comparisons between vegetation conditions. Blue, red, and near-infrared reflectances, centered at 469, 645, and 858 nm, respectively, are used to determine the MODIS daily vegetation indices. Global MOD13A3 data are provided as monthly average at 1-km spatial resolution

MCD43B3

The MODIS *albedo* product provides 1 km data describing both directional hemispherical reflectance (*black-sky albedo*) at local solar noon and bihemispherical reflectance (*white-sky albedo*). Measurements are averaged over 16 days. Both the near-infrared albedo and visible albedo have been acquired for this project
MOD44 W: MODIS 250 m land–water mask is a land cover–based global land–water mask

MOD44 W

MODIS 250 m *land–water mask* is a land cover–based global land–water mask

freely available via the website[1] of LP DAAC,[2] those in (Table 4.2) have been acquired.

Samples of MODIS products were acquired in *HDF-EOS* format and have been submitted to preprocessing modules including tiling, geocoding, and resampling. These functions are implemented by a dedicated software program, the MODIS *Reprojection Tool* (MRT).[3] The images taken by the MODIS were collected in the same period to ensure that they were comparable. In addition, to capture the *maximum thermal excursion* in the region, July was chosen as the reference period for the images. To carry out the lithological classification of surfaces with the satellite images, a limited number of control points have also been used.

---

[1] https://lpdaac.usgs.gov/products/modis_products_table

[2] NASA Land Processes Distributed Active Archive Center (LP DAAC). MODIS products. USGS/Earth Resources Observation and Science (EROS) Center, Sioux Falls, South Dakota. 2012.

[3] https://lpdaac.usgs.gov/tools/modis_reprojection_tool

Table 4.3 CORINE categories data acquired

CORINE category code	CORINE category description	Main purpose
111	Continuous urban fabric	Urban mask
112	Discontinuous urban fabric	Urban mask
311	Broad-leaved forest	Forest mask
312	Coniferous forest	Forest mask
313	Mixed forest	Forest mask

4.2 Vegetation Coverage and Land Use

A satellite image represents the emissivity of the shallowest layer of ground and is indicative of the rock properties only in the sites which are not covered by dense vegetation. The pixels corresponding to sites with dense vegetation, such as forests, have been masked to avoid that they jeopardize the classification process. The same applies for the pixels falling in urban areas.

To build the masks, data acquired by the European project *Coordination of Information on the Environment* (CORINE) Land Cover (CLC) have been used. The aim of CORINE project is the provision of a unique and comparable dataset of land cover for Europe. The mapping of the land cover and land use, performed on the basis of satellite remote sensing images on a scale of 1:100,000, provides land use classification comprising 44 classes. The data acquired for this work are extracted from the 2006 update project CLC2006. CORINE data are freely available from the website of the *European Environmental Agency,*[4] from which the shapefiles corresponding to the categories listed in Table 4.3 have been obtained. The shapefiles have been combined into two distinct mask layers: an *urban coverage mask* and a *forest coverage mask* and then saved in *GeoTiff* file format.

4.3 Energy Demand Data

The energy needs for thermoregulation and the geothermal potential determine the size of the GSHP plant. The energy need for heating is conventionally expressed in *degree*-days (celsius*day/year), calculated as the sum of daily positive differences between the reference temperature of the home, in Italy conventionally set at 20 °C, and the average daily outdoor temperature. This sum is extended over the whole year. In Italy, the value of *Degree-Days* (DD) is fixed for each municipality by national regulations (Repubblica Italiana 1993), and the municipality is also assigned to a climate zone on a conventional scale based on this value. Each

---

[4] http://www.eea.europa.eu

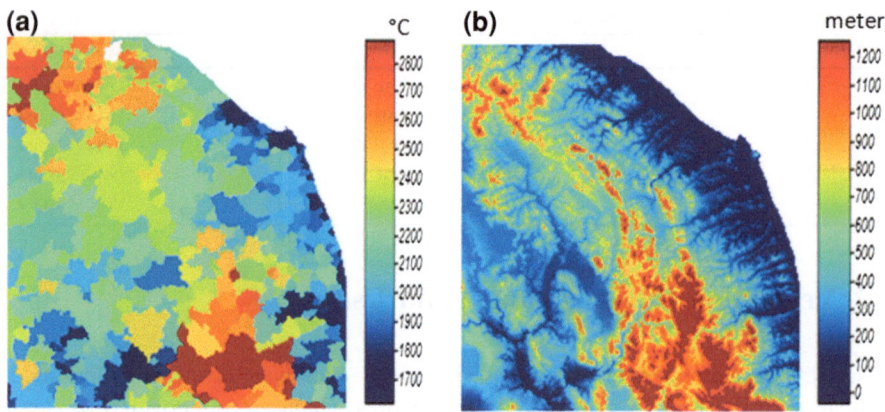

Fig. 4.1 a Degree-days; **b** digital terrain model

European country has issued similar regulations. Figure 4.1a presents the layer of degree-days.

The depth of the BHE is determined by the local *specific heat extraction* and the energy needs. A depth correction factor is also added to consider the surface temperature of the soil (GST). To estimate this correction factor, a GIS layer for the GST feature has been produced. Two types of data were used to this end:

- *Mean data of Surface Air Temperature* (SAT) originated by several meteorological stations scattered throughout the study region, within the period 1950–2000 and acquired from various sites freely accessible on the World Wide Web.[5,6].
- The *Digital Terrain Model* (DTM) layer (Fig. 4.1b) is a raster product from the *NASA Shuttle Radar Topography Mission* (SRTM) and was acquired from (Jarvis et al. 2009) with a spatial resolution of 90 m, in *ArcInfo ASCII* format.

The procedure to obtain the GST layer is as follows:

1. Calculate the thermal gradient elevation across the region with a polynomial regression between the elevation data and measures of SAT.
2. Normalize the SAT measures at sea level using the quotas from DTM and regression model.
3. Geostatistical interpolation (*Kriging*) of normalized SAT.
4. Bring back to the original elevation the interpolated temperatures using the regression model. The product obtained is a GIS layer of the SAT.
5. Subtract 1.6 °C from the SAT to get the GST (Fig. 4.2a).

---

[5] Centro di Ecologia e Climatologia Marche, Osservatorio Geofisico Sperimentale di Macerata, http://www.geofisico.it/

[6] Atlante Climatico d'Italia by Servizio Meteorologico dell'Aeronautica Militare http://clima.meteoam.it/AtlanteClimatico/Index.htm.

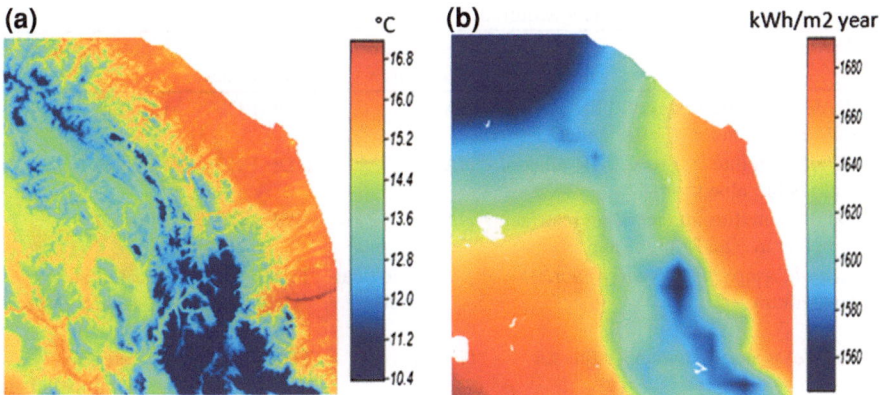

Fig. 4.2 a Ground surface temperature, **b** global solar irradiation

The *solar irradiation* has to be included in the analysis to set the size of solar thermal plants that can improve the efficiency of a BHE plant. The solar irradiation dataset has been acquired from the database of the *Photovoltaic Geographical Information System* (PVGIS)[7] and represents a collection of average annual irradiation (kWh/m^2) measurements (Suri et al. 2007). In particular, the acquired irradiation data correspond to collectors with optimal surface exposure to the sun. A continuous layer of radiation was obtained (see Fig. 4.2b).

4.4 Summary of the Data Acquired

Table 4.4 shows the complete list of the acquired data along with the corresponding source, acquisition format, and original spatial resolution.

Each GIS layer has been brought to a common data structure with the following procedure:

1. *Resampling*: the continuous surface in the study region was divided into a grid of 1,818 rows and 1,670 columns. The resulting cells are 30,36,060, each one representing a geographic area of 100 × 100 m.
2. *Geographic reprojection* for all layers has been adopted the projected coordinate system UTM Zone 33, datum WGS84.
3. *Export to GeoTIFF* raster format that can be read by GIS software.

---

[7] http://re.jrc.ec.europa.eu/pvgis/

Table 4.4 Summary table of data acquired

Product name/description	Organization data source	Acquisition data format	Input resolution
MOD13A3/satellite imagery, blue reflectance, and red reflectance	NASA	HDF-EOF	1,000 m
MCD43B3/satellite imagery, NIR albedo, and visible albedo	NASA	HDF-EOF	1,000 m
MOD11A2/satellite imagery, land surface temperature (LST) night and day	NASA	HDF-EOF	1,000 m
Surface air temperature/yearly average temperature	Meteorological services	TEXT	Located at meteorological stations
Degree-days/energy demand indicator	Normatives	TEXT	Referred to administrative unit
Digital terrain model	CGIAR consortium	ARCInfo ASCII	90 m
Global solar irradiation	JRC	ARCInfo ASCII	1,000 m
Vegetation and urban masks	EEA	Shapefile	N/A
Geologic ground truth	Direct survey	TEXT	Sparse points
Commune boundary	ISTAT	Shapefile	

Table 4.5 Summary table of geographic features acquired

Layer	Units
LST night	°C
LST day	°C
Red reflectance	Normalized
Blue reflectance	Normalized
Albedo visible	Normalized
Albedo NIR	Normalized
Commune boundaries	N/A
Water mask	Boolean
Urban mask	Boolean
Forest mask	Boolean
Degrees-day	°C* day/year
Digital terrain model	meters
Ground surface temp.	°C
Solar radiation	kWh/m2

In Table 4.5 are shown the GIS layers made available at the completion of preprocessing work.

References

Jarvis A, Reuter HI, Nelson A, Guevara E (2009) Hole-filled seamless SRTM data V4. International Centre for Tropical Agriculture (CIAT)

Repubblica Italiana (1993) D.P.R. 26 agosto 1993, n. 412. Gazzetta Ufficiale 1993[214]. Gazzetta Ufficiale. Serie Generale. Ref Type: Statute

Suri M, Huld TA, Dunlop ED, Ossenbrink HA (2007) Potential of solar electricity generation in the European Union. Sol Energy 81:1295–1305

References

Chapter 5
Feature Analysis: Selecting Decision Criteria

Abstract The data acquired constitute a wealth of information still to be reduced and transformed for the specific goals of the decision problem. In this chapter, the automated reasoning techniques are used to maximize the efficiency of data processing, and to extract indicators of the LTGE potential and of the energy demand, this resource must meet. The final products of this process are indicators of performance of the LTGE plant and other alternative systems. The cost-benefit comparison of the plants is referred to a site chosen as reference and to standardized production.

The purpose of feature analysis in this case study is to apply feature selection to (1) improve the classification process and (2) compute derived features that are not available among the acquired data but are necessary to complete the decision process. The feature analysis leads to a set of features that are subsequently used as decision criteria in this case study.

The feature analysis is divided into three phases:

1. Processing the satellite images to produce a GIS layer that represents the potential extractable geothermal resource.
2. Production of layers representing the size and cost of the GSHP plant.
3. Production of layers representing the cost of alternative systems.

5.1 Satellite Image Processing: Geothermal Potential Assessment

This section describes the process that leads to a spatial model of the natural geothermal potential. As mentioned earlier, the *Specific Heat Extraction* (sHE) of the soil represents an effective measure of geothermal potential. In each site, the sHE depends on the type of rock lying in the underground and can be estimated if the dominant lithology of the site is known. Therefore, the problem of constructing

A. Gemelli et al., *GIS to Support Cost-Effective Decisions on Renewable Sources*, SpringerBriefs in Applied Sciences and Technology, DOI: 10.1007/978-1-4471-5055-8_5, © The Author(s) 2013

a GIS layer of sHE potential is equivalent to the problem of constructing a GIS layer of the lithology. The recognition of lithology from satellite imagery is particularly a difficult task due to the variety of geographic objects such as vegetation, debris, and construction that hinder the direct view of the rock constituting the storage reservoir. However, as will be shown, the computational approach to the analysis of satellite images leads to accurate results despite the uncertainty due to disturbance factors. The overall process that leads to the model of natural geothermal potential is described in Fig. 5.1.

The features available for classification process are as follows:

1. *blue reflectance*;
2. *red reflectance*;
3. *albedo visible*;
4. *albedo NIR*;
5. *apparent thermal inertia*;
6. *slope*.

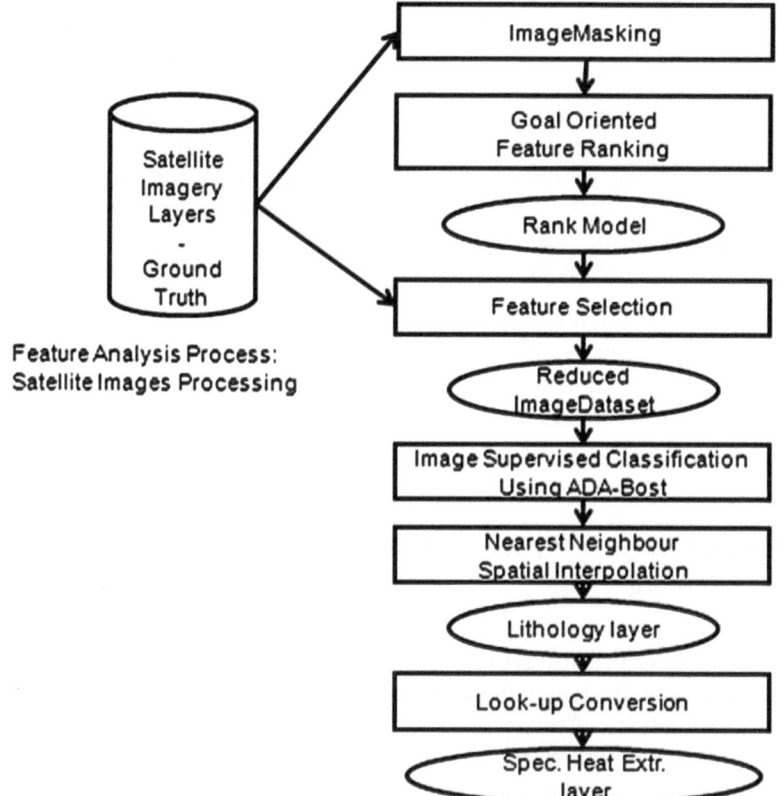

Fig. 5.1 Lithological mapping process

where

- the first four features correspond to layers acquired directly from satellite images (Table 4.5);
- the layer of apparent thermal inertia is obtained from LST—Night, LST—day, and albedo NIR images, using the Eq. 1.3;
- the layer of slope is obtained from the DTM. In it, the slope has been discretized into 10 intervals. In the geologic survey of a wide area, the slope is considered to be a highly discriminative feature of the underground rock type.

In each raster, the cells corresponding to sites covered by water (sea, lake), vegetation, and urban areas have been masked. The unmasked cells are of 1436241 (47.3 % of the whole region).

In the *geocomputation* approach, each satellite image product represents a feature that hypothetically contributes to achieving the goals of the lithological classification. However, the value of this contribution is not known and presumably differs among the images. To measure the individual importance of each feature, we apply the Goal-Oriented Feature Ranking (GOFR). To this end, a *training subset* of 1,577 records has been extracted with its corresponding lithological class. The *training subset* represents the ground truth data. The lithological classes are as follows: *limestone, gravel sand dry, gravel sand water saturated, clay marls, clay,* and *sandstone*, based on the classification of (Verein Deutscher Ingenieure 2001), which aimed to distinguish rocks based on their thermal characteristics.

5.1.1 Applying the Goal-Oriented Feature Ranking Method

The training subset has undergone the following feature selection procedure:

- Application of GOFR using OLDA as core FE algorithm.
- The features are ordered by the ranks calculated with GOFR.
- A series of subsets has been constructed. The first subset contains only the feature with the highest weight, and the following five subsets are obtained by progressively adding a feature, picking the one with the highest weight among the features left. The sixth subset contains six features.
- The *ADABoost* algorithm has been applied separately to each subset using the 10-fold cross-validation scheme, and the accuracy of classification is calculated. At the end of the validation, 6 accuracy values have been obtained, showing progress in the quality of classification when adding features with progressively lower weights.

By applying the 10-fold cross-validation, the training set has been divided 10-fold. The weights of the features and the accuracy of the classification are reported as the averages of the results obtained for each fold. This information was used to construct the model ranking in Fig. 5.2a. The whole procedure was also applied with other classification algorithms: LVQ and discriminant analysis. The *curves of accuracy* of the three classifiers are compared in Fig. 5.2b.

Before proceeding with feature selection, it has been observed that (referring to Fig. 5.2b)

- ADABoost achieves accuracy rates well above those of the other algorithms;
- The accuracy of ADABoost grows monotonically and reaches a value close to 100 % with only four features, leading us to consider the two remaining features as useless. In contrast, the accuracy of discriminant analysis shows an intermediate peak on the second feature, beyond which the accuracy tends to decrease, similarly for the LVQ, for which there is a peak value on the fourth feature. These effects are due to curse of dimensionality.

It possible to assert that in this case study, the problem of recognition of lithology is addressed in an optimal way with the choice of ADABoost as the classification algorithm, working on a dataset reduced to slope, visible albedo, albedo NIR, thermal inertia.

5.1.2 Layer of Lithology: Applying ADA-Boost Classification

ADABoost is applied to the dataset of the whole region and trained with the set of 1,577 labeled records from the ground truth data. Both the test and training sets only contain the four selected features. The 1436241 unmasked points are labeled according to the extracted model. ADABoost was performed in 35 iterations.

For the masked sites, the lithology has been obtained with a *spatial nearest-neighbor interpolation*. The lithology layer is shown in (Fig. 5.3). The histograms show that the dominant lithology is clay, followed by limestone and gravel sand.

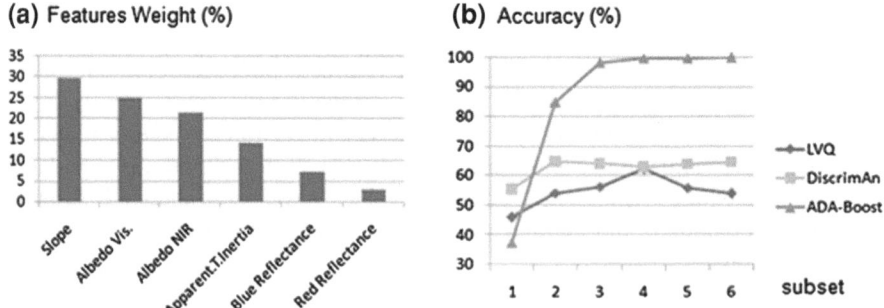

Fig. 5.2 a Feature ranking model, **b** classification accuracy curves

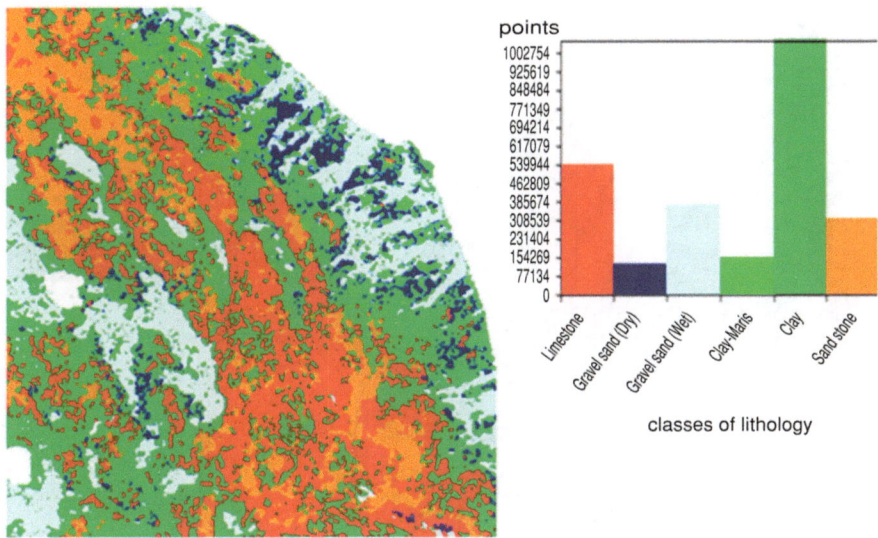

Fig. 5.3 Lithology layer and class distribution

5.1.3 Specific Heat Extraction Layer

The layer of the sHE (Fig. 5.4) is obtained directly from the lithology by converting each class of lithology to a value of sHE using Table 5.1. This table is extracted from the (Verein Deutscher Ingenieure 2001), which can be considered as a lookup table.

5.2 Dimensioning and Cost Evaluation of Geothermal Plants

A class "E" house according to the *Climate-House* energy efficiency classification (European Community 2002) (Lantschner 2005) consumes 120 kWh/m^2 per year for domestic heating. For each climate zone, the average annual energy consumption for heating a *class* "E" house was acquired from (Santini et al. 2009). Then, a linear regression was performed to find a relationship that relates the energy needs H_u to the degree-day indicator:

$$H_u \left(\text{kWh m}^{-2} \text{ year}^{-1} \right) = \text{DD} \cdot 0.0411 - 4.677$$

The energy needs are then multiplied by 100 m^2 (the average surface of a house). The result is the *yearly energy requirement for heating* a 100 m^2 class E house (Fig. 7.1a):

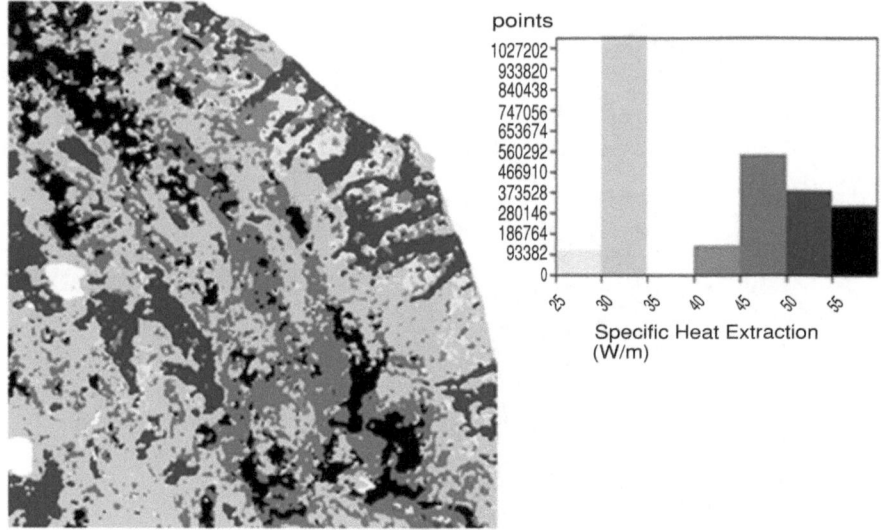

points

Specific Heat Extraction
(W/m)

Fig. 5.4 Specific heat extraction layer and class distribution

$$H\left(\text{kWh year}^{-1}\right) = H_u\left(\text{kWh m}^{-2}\,\text{year}^{-1}\right) * 100(\text{m}^2)$$

Taking the site corresponding to the locality Fabriano (Fig. 7.1a) as an example, and for a 100 m^2 class E house, we have

$$H_u = 95 \text{ kWh m}^{-2}$$

and

$$H\left(\text{kWh year}^{-1}\right) = 95\left(\text{kWh m}^{-2}\text{year}^{-1}\right) * 100\left(\text{m}^2\right) = 9500\left(\text{kWh year}^{-1}\right)$$

Note that the yearly energy for heat extracted by the BHE is $H_{\text{BHE}} < H$ because the electricity H_{EL} consumed by the GSHP is directly converted into heat and satisfies a portion of the need. Therefore, the following conditions exist:

$$H_{\text{BHE}} = H - H_{\text{EL}}$$

Table 5.1 Specific heat extraction lookup table

Class#	Class of lithology	Specific heat extraction (W/m)
1	Limestone	45
2	Gravel, sand, dry	25
3	Gravel, sand, water saturated	50
4	Clay marls	40
5	Clay	30
6	Sandstone	55

$$H_{EL} = H_{BHE}/COP$$

The same GSHP plant can also be used for cooling. Assuming that E_{BHE}, the *yearly energy need for cooling* a $100\ m^2$ class E house is $E = 0.83 *$ $H = 7{,}885.9\ kWh\ year^{-1}$ and $EER = 14$, the following conditions exist:

$$E_{BHE} = E - E_{EL}$$

$$E_{EL} = E_{BHE}/EER$$

Calculations satisfying all the conditions indicate that $H_{EL} = 2{,}111.1\ kWh$ $year^{-1}$, $E_{EL} = 525.67\ kWh$ $year^{-1}$, $H_{BHE} = 7{,}388.9\ kWh$ $year^{-1}$, $E_{BHE} = 7{,}360.23\ kWh\ year^{-1}$, and assuming a pump operating time of 2,400 h/ year for both heating and cooling operations, the BHE must provide a power of about $P_{BHE} = 3{,}078$ watts. Since the site has a $sHE_{agg} = 35.4\ Wm^{-1}$, the depth that the BHE must reach is calculated by applying Eq. 1.5:

$$Z_{BHE'} = 3{,}078\ (W)/35.4\ \left(Wm^{-1}\right) + 0.64\ (m) = 87.58\ (m)$$

Based on real price lists, the components of a GSHP plant have costs as shown in Table 5.2. At the Fabriano site, the up-front cost of a GSHP plant ($C_{GSHP/front}$) is

$$C_{GSHP/front} = 15{,}132.8\,€$$

The maintenance costs of the GSHP plant, $C_{GSHP/year}$, are due only to the cost of the electricity consumed by the heat pump. Hence, assuming an electricity cost of $0.18\ €\ kWh^{-1}$, it is possible to determine that the yearly cost of maintenance of the whole system is

$$
\begin{aligned}
C_{GSHP/year}(€) &= (H_{EL} + E_{EL}) * 0.18\,€\ kWh^{-1} \\
&= (2{,}111.1 + 525.67)\ kWh\ year^{-1} * 0.18\,€\ kWh^{-1} \\
&= 474.62\,€\ year^{-1}
\end{aligned}
$$

The consumption of electricity is the only source of pollution in a GSHP plant. The yearly CO_2 equivalent emission by the GSHP plant is

$$
\begin{aligned}
CO_2\ Equiv_{GSHP/year} &= (H_{EL} + E_{EL}) * EF_{EL} \\
&= (2{,}111.1 + 525.67)\ kWh\ year^{-1} * 0.43\ Kg\ CO_2\ kWh^{-1} \\
&= 1{,}133.81\ Kg\ CO_2
\end{aligned}
$$

Table 5.2 Up-front costs of GSHP

GSHP components	Cost
Drilling	87.58 m* €/m
Tubes	87.58 m* 10 €/m
Heat pump	2,480.0 €
Radiators	6,100.0 €
Boiler 500 L	1,298.0 €

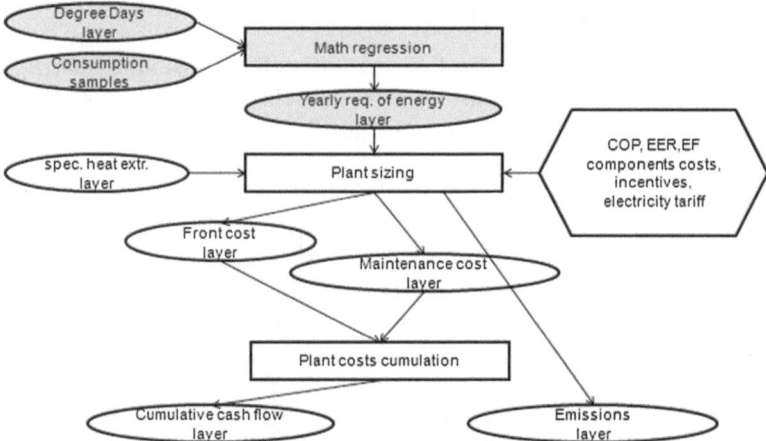

Fig. 5.5 Procedure used to calculate the plant costs

These calculations for the Fabriano site have been applied systematically to all sites in the region, and the resulting layers are reported in the atlas section of this book: Fig. 7.1 a) yearly energy needs (H) for heating a class E house of 100 m^2. Figure 7.1b $C_{GSHP/front}$, up-front cost of GSHP plant. Figure 7.2c $C_{GSHP/year}$ GSHP yearly maintenance cost. Figure 7.4c CO_2 $Equiv_{GSHP/year}$ emissions. In all figures, the reference site Fabriano is marked with a tiny circle.

The process used to calculate the size of the GSHP is shown in Fig. 5.5. Rectangles are the process steps, the ovals represent the GIS layer I/O to the process, and the hexagon represents a set of fixed parameters such as the electricity tariff.

The following sections analyze the costs of plants other than the GSHP. Two applications are considered: (1) plants dedicated to thermoregulation and (2) plants integrating thermoregulation and domestic hot water (DHW) production. The analysis is still conducted on the Fabriano reference site and can then be extended later to every site in the region. The analysis scheme in Fig. 5.5 is extended to all types of plants. Note that all the plants are sized in a way to provide the same yearly energy needs for heating and that this calculation is performed only once (shaded elements of Fig. 5.5).

5.3 Alternative Plants for the Thermoregulation

5.3.1 Hybrid System Methane and Split

In this section, the cost of a hybrid plant composed of a methane subsystem for heating and Split subsystem for cooling is calculated. The abbreviation METsplit (see the Front Matter for a complete *list of abbreviations*) is adopted to refer to this type of system.

Based on real price lists, the components of a *METsplit* plant have costs as shown in Table 5.3. The sum is the up-front cost of a *METsplit* plant, $C_{METsplit/front}$ At the Fabriano site:

$$C_{METsplit/front} = 7,350.0 \,€$$

The maintenance costs of the plant are calculated in the following way.

Given the yearly needs of energy (for heating), $H = 9,500$ kWh year^{-1}, the cost of methane for heating is

$$\text{Yearly cost of energy (for heating)} = 9,500 \text{ kWh year}^{-1}/((\text{calorific value of natural gas})$$
$$9.6 \text{ kWh m}^{-3} * (\text{efficiency}) \, 0.91) * 0.84 \,€\text{m}^{-3}$$
$$= 913.46 \,€ \text{ year}^{-1}$$

Given the yearly needs of energy (for cooling), $E = 7,885.90$ kWh year^{-1}, the cost of electricity to operate the heat pump of the Split (with $EER_{split} = 3.15$) is

$$\text{Yearly cost of energy (for cooling)} = 7,885.9 \text{ kWh year}^{-1}/3.15 * 0.18 \,€ \text{ kWh}^{-1}$$
$$= 450.62 \,€\text{year}^{-1}$$

For support activities (cleaning, hygienization, charging), a cost of approximately 320 € year^{-1} must be added. Hence, the yearly overall costs for maintenance of the *METsplit* plant are

$$C_{METsplit/year} = 913.46 \,€ \text{ year}^{-1} + 450.57 \,€ \text{ year}^{-1} + 320 \,€ \text{ year}^{-1}$$
$$= 1,684.03 \,€ \text{ year}^{-1}$$

The *METsplit* plant has polluting emissions due to consumption of electricity and methane. The yearly CO_2 equivalent emission by this plant is

$$CO_2 \text{ Equiv}_{METsplit/year} = H * EF_{ME} + E/EER_{split} * EF_{EL}$$
$$= 9,500 * 0.19 + 7,885.9/3.15 * 0.43 \text{ Kg } CO_2 \text{ kWh}^{-1}$$
$$= 2,881.48 \text{ Kg } CO_2$$

The cash flows of METsplit and GSHP plants are compared, assuming that a government *incentive plan* is active in the region. The economic incentives affect

Table 5.3 Up-front cost of components of mixed system methane and Split	Plant: methane + Split	Cost
	Condensing boiler	2,000.0 €
	Split (COP = 3.5, EER = 3.15) 4 kW	1,350.0 €
	Radiators	4,000.0 €

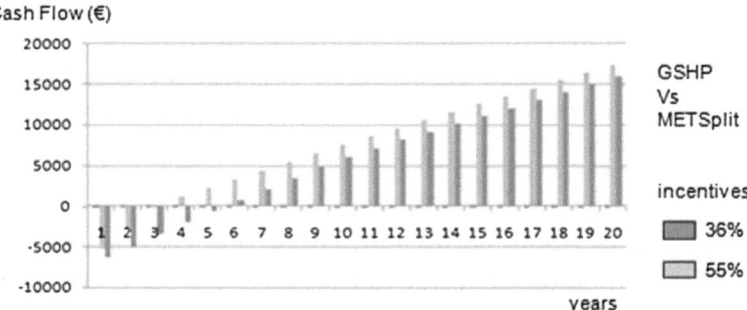

Fig. 5.6 Cumulative savings of GSHP over METsplit, for 36 and 55 % incentives

significantly the construction and maintenance costs of the plants and their convenience. To highlight the effect of incentives, the examples that follow two alternative incentives (actually available in Italy) are applied:

- a discount on the taxes, equivalent to 36 % of the up-front cost of the plant, to be taken within 10 years;
- a discount on the taxes, equivalent to 55 % of up-front cost of the plant, to be taken within 3 years.

The savings through the years of the GSHP plant over the METsplit are shown in Fig. 5.6. The savings are calculated with the following procedure:

1. Up-front cost of the METsplit plant: the yearly maintenance costs, and the 36 % incentives, have been accumulated for 20 years for the METsplit plant
2. Up-front cost of the GSHP plant: the yearly maintenance costs, and the 36 % incentives, have been accumulated for 20 years for the GSHP plant
3. the difference 1–2 is calculated.

In Fig. 5.6, the intersection of trends with the time axis indicates that the payback period of a GSHP plant relative to the METsplit is 5–6 years with 36 % incentives and 3 year with 55 % incentives.

5.3.2 Split Plant

The cost of a thermoregulation plant that is composed only of a Split system for heating and cooling is calculated in this section. The name Split is adopted to refer to this plant.

The up-front cost, $C_{\text{Split/front}}$, of the Split plant is shown in Table 5.4 and is extracted from a real price lists:

$$C_{\text{Split/front}} = 1,749.0 \, €$$

Split plant	Cost
Split (COP = 3.5, EER = 3.15) 5 kW	1,749.0 €

Table 5.4 Up-front costs of Split plant

Given the yearly needs of energy (for cooling), $H = 9,500$ kWh year^{-1} :

Cost of electricity for the Split operating in heating mode

$$= 9,500.0\,\text{kWh}\ \text{year}^{-1}/(\text{COP}_{\text{split}})3.5 * 0.18\text{€ kWh}^{-1}$$

Given the yearly needs of energy (for cooling), $E = 7,885.90$ kWh/year:

Cost of electricity for the Split operating in cooling mode

$$= 7,885.9\,\text{kWh}\ \text{year}^{-1}/(\text{EER}_{\text{split}})3.15 * 0.18\text{€ kWh}^{-1} = 450.62\text{€ year}^{-1}$$

For support activities (cleaning, hygienization, charging), a cost of about 100 € year^{-1} must to be added. The yearly overall costs for maintenance of the Split plant are

$$C_{\text{Split/year}} = 488.57\,\text{€ year}^{-1} + 450.57\,\text{€ year}^{-1} + \text{€ year}^{-1}$$
$$= 1,039.14\,\text{€ year}^{-1}$$

The Split plant has emissions due only to consumption of electricity. The yearly CO_2 equivalent emission by the plant is

$$CO_2\ \text{Equiv}_{\text{Split/year}} = H/\text{COP}_{\text{split}} * \text{EF}_{\text{EL}} + E/\text{EER}_{\text{split}} * \text{EF}_{\text{EL}} = 9,500/3.5 * 0.43$$
$$+ 7,885.9/3.15 * 0.43\ \text{Kg CO}_2\text{kWh}^{-1}$$
$$= 2,243.63\ \text{Kg CO}_2$$

As in the previous case, the cash flows of Split and GSHP plants are compared, assuming that a government incentive plan is active in the region. In Fig. 5.7, the savings through the years of the GSHP plant over the Split are shown. The intersection of trends with the time axis indicates that the payback period of the GSHP plant relative to the Split is 18 years with 36 % incentives and 12 years with 55 % incentives. The cost savings of GSHP plant versus Split is weaker than GSHP versus METsplit. However, it should be considered that the GSHP plant has a life expectancy of 20 years, whereas both the Split and METsplit plants require a full renovation every 10 years causing additional costs.

5.4 Integration of Thermoregulation with DHW Plant

In this section, the costs of integrated plants of thermoregulation and DHW are calculated.

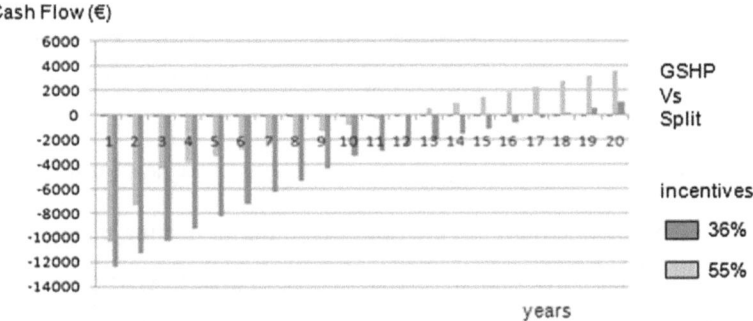

Fig. 5.7 Cumulative savings of GSHP over Split, for 36 and 55 % incentives

5.4.1 Integrated Plant: GSHP and Solar Hot Water

The integration of the solar hot water (SHW) plant with the GSHP plant is the most effective way to combine solar power with geothermal power, and it is already a widely adopted solution. In this case study, the integration is achieved by oversizing the SHW plant to satisfy a part of the energy demand for thermoregulation. Consequently, the GSHP plant is downsized. As will be seen, this leads to a significant reduction in the overall cost. In the rest of this section, the abbreviation GSHP&SHW is adopted to refer to this type of integrated plant.

A SHW plant of the *forced circulation* type is considered. With reference to the site of the Fabriano, the yearly energy needs for the DHW, H_{DHW} are calculated for a house of 100 m^2, assuming that there is an inhabitant each 36.8 m^2, as it is in the national average:

$$H_{\mathrm{DHW}} = \text{inhabitants} * \text{liters\_for\_inhabitant} * \text{water\_specific\_heat} *$$
$$(\text{DHW\_temperature} - \text{environment\_water\_temperature})/1,000 * \text{Days}$$
$$= 2.718 * 70 * 1.16 * (45 - 14.3)/1,000 * 365 = 2,473.07\,\mathrm{kWh}\ \mathrm{year}^{-1}$$

The water temperature is fixed to 45 °C for sanitary requirements, and the *environment_water_temperature* corresponds to Ground Surface Temperature (GST) in the site.

The SHW plant is oversized to produce an amount of heat equivalent to $^1/_2 H_{\mathrm{BHE}}$; in addition to H_{DHW}, it provides for the hot water already. Therefore, the SHW plant must meet the following energy needs:

$$H_{\mathrm{DHW+}} = H_{\mathrm{DHW}} + ^1/_2 H_{\mathrm{BHE}}$$
$$= 2,473.07\ \mathrm{kWh}\ \mathrm{year}^{-1} + ^1/_2\,7,388.9\ \mathrm{kWh}\ \mathrm{year}^{-1}$$
$$= 6,167.52\ \mathrm{kWh}\ \mathrm{year}^{-1}$$

Table 5.5 Up-front costs of SHW plant

SHW plant	Cost
Solar panel cost	220.0 €/m^2
Boiler 500 L	1,298.0 €
Pump, pipes, heat exchanger, anti-cooling liquid, valves	600.0 €
Electronic control unit	137.0 €

Based on real price lists, the components of a SHW plant have costs as shown in Table 5.5, based on real price lists.

The up-front cost of the SHW plant is calculated according to the energy demand:

$$
\begin{aligned}
C_{\text{SHW/front}} &= \text{surface\_of\_the\_solar\_panels} * \text{specific\_cost\_of\_panels} + \text{other\_costs} \\
&= H_{\text{DHW+}} * \text{solar\_coverage\_factor/global\_solar\_irradiation/efficiency}) * \\
&\quad \text{specific\_cost\_of\_panels} + \text{extra\_costs} \\
&= \left(6,167.52\,\text{kWh year}^{-1} * 0.65/1,615.11\,\text{kWh year}^{-1}/0.5\right) * 220\,€\text{m}^2 \\
&\quad + 2,035€ = 3,127.13€
\end{aligned}
$$

The global solar irradiation is taken from the layer in Fig. 4.2b. At the Fabriano site: global solar irradiation = 1,615.11 kWh year^{-1}.

After satisfying half of the energy needs for heating with SHW plant, the energy production by BHE can be halved and so its depth. The up-front cost of GSHP subsystem resized is obtained from Table 5.2, by halving the depth of BHE, and it is 12,504.4 €.

The overall up-front cost of GSHP&SHW plant ($C_{\text{GSHP\&SHW/front}}$) is

$$
C_{\text{GSHP\&SHW/front}} = 3,127.13 + 12,504.4 = 1,5631.53€.
$$

It should be noted that the cost of the two plants is not integrated: (1) the SHW dedicated only to the production of hot water and (2) GSHP only for thermoregulation would be: 2,164.0 € + 15,132.8 € = 17,296.8 €, which is higher than the integrated plant.

With regard to the GSHP&SHW plant, also the power consumption by GSHP is halved: $^1\!/_2\,C_{\text{GSHP/year}} = €\;237.31$ year^{-1}. Therefore, assuming that the power consumption of the SHW pump is $E_{\text{SHW}} = 555$ kWh year^{-1}, equivalent to the cost of 99.9 € year^{-1}, the yearly overall maintenance costs of the GSHP&SHW plant are

$$
C_{\text{GSHP\&SHW/year}} = 237.31€ = \text{ year}^{-1} + 99.9€ = \text{year}^{-1} 337.21€ \text{ year}^{-1}
$$

The GSHP&SHW plant has emissions due to consumption of electricity by GSHP and by the pump of the SHW subsystem. The yearly CO_2 equivalent emission by the GSHP&SHW plant is

$$
\begin{aligned}
CO_2 \text{ Equiv}_{\text{GSHP\&SHW/year}} &= ((H_{\text{EL}} + E_{\text{EL}})/2 + E_{\text{SHW}}) * \text{EF}_{\text{EL}} \\
&= ((2,111.1 + 525.67)/2 + 555) * 0.43 \text{ Kg CO}_2/\text{kWh} \\
&= 805.55 \text{ Kg CO}_2
\end{aligned}
$$

5.4.2 Integrated Plant: Methane for DHW and Split for Thermoregulation

An integrated plant with a methane subsystem for DHW and a Split subsystem for thermoregulation, hereafter indicated by the abbreviation Split&DHW, is analyzed. For the equivalence of the components, we assume the up-front cost Split&DHW is equal to that of Table 5.3:

$$C_{\text{Split\&DHW/front}} = 7350.0 \, \text{€}$$

The maintenance costs of the Split&DHW plant is calculated in this way: for DHW, as in the previous calculation, is required $H_{\text{DHW}} = 2,473.07$ kWh year^{-1}

Maintenance cost of the methane subsystem for DHW

$$
\begin{aligned}
&= 2,473.07 \, \text{kWh year}^{-1}/((\text{methane\_calorific\_value}) \, 9.6 \, \text{kWh m}^{-3} * (\text{efficiency}) \, 0.91) \\
&\quad * 0.84 \, \text{€ m}^{-3} \\
&= 237.79 \, \text{€ year}^{-1}
\end{aligned}
$$

Maintenance costs of the subsystem Split $= C_{\text{Split/year}} = 1,039.14 \, \text{€/year}$

Adding the costs of support of the methane subsystem, 220 €/year, the yearly overall cost of maintenance of the Split&DHW plant is

$$
\begin{aligned}
C_{\text{Split\&DHW/year}} &= 237.79\text{€/year} + 1,039.14\text{€/year} + 220\,\text{€/year} \\
&= 1,496.93
\end{aligned}
$$

The Split&DHW plant has emissions due to consumption of electricity and methane. The yearly CO_2 equivalent emission by the Split&DHW plant is

$$
\begin{aligned}
CO_2 \text{ Equiv}_{\text{Split\&DHW/year}} &= (H/\text{COP}_{\text{split}} + E/\text{EER}_{\text{split}}) * \text{EF}_{\text{EL}} + H_{\text{DHW}} * \text{EF}_{\text{ME}} \\
&= (9,500/3.5 + 7,885.9/3.14) * 0.43 + 2,473.07 * 0.19 \text{ Kg CO}_2 \text{ kWh}^{-1} \\
&= 2,716.94 \text{ Kg CO}_2
\end{aligned}
$$

5.4.3 Integrated Plant: Methane for DHW/Heating and Split for Cooling

An integrated system composed of a methane subsystem for thermoregulation and DHW and Split subsystem for air cooling, hereafter indicated by the abbreviation *MET&DHW*, is analyzed.

For the equivalence of the components, we assume that the up-front cost of the plant is equal to that of Table 5.3; therefore,

$$C_{\text{MET\&DHW/front}} = 7,350.0\,€$$

The maintenance costs of the methane subsystem are so calculated: For the DHW, as in the previous sections, is needed $H_{\text{DHW}} = 2473.07$ kWh year^{-1} and given the yearly needs of energy (for heating) $H = 9,500$ kWh/year:

Cost of maintenance of methane subsystem for DHW and heating

$= (2,473.07 + 9,500\,\text{kWh year}^{-1})/$

$((\text{calorific value of met.})9.6\text{ kWh m}^{-3} * (\text{performance})\,0.91) * 0.84€\text{ m}^{-3}$

$= 1,151.25€\text{ year}^{-1}$

For a given yearly needs of energy (for cooling) is $E = 7,885.90$ kWh year^{-1} , the maintenance costs of the Split subsystem, is:

$= 7,885.9\text{ kWh year}^{-1}/(\text{EER}_{\text{split}})3.15 * 0.18\,€nbsp; \text{ kWh}^{-1} = 450.62\,€\,\text{year}^{-1}$

Adding the costs of 320 € year^{-1}/year for support activities, the yearly overall cost of maintenance of the MET&DHW plant is

$$C_{\text{MET\&DHW/year}} = 1,151.25€/\text{year} + 450.62€/\text{year} + 320\,€/\text{year}$$
$$= 1,921.87€/\text{year}$$

The MET&DHW plant has emissions due to consumption of electricity and methane. The yearly CO_2 equivalent emission by this plant is

$$C_{\text{MET\&DHW/year}} = 1,151.25€/\text{year} + 450.62€/\text{year} + 320\,€/\text{year}$$
$$= 1,921.87€/\text{year}$$

5.5 Comparison of Costs and Benefits of Plants

The comparative analysis of the costs involved in the Fabriano site, presented in Table 5.6, demonstrates that LTGE plants are approximately twice the cost of other plants, even though they satisfy the same energy needs. However, the yearly maintenance costs of GSHP plants are less than 50 % compared with other plants.

Table 5.6 Comparative table of costs and benefits of plants

Plant ref. name	Heating subsystem	Cooling subsys.	DHW subsys.	Front cost (€)	Maintenance cost	Emissions
GSHP	GSHP	GSHP	–	15,132.8	474.6	1,133.8
METsplit	Methane	Split	–	7,350.0	1,684.0	2,881.5
Split	Split	Split		1,749.0	1,039.1	2,243.6
GSHP&SHW	GSHP&SHW	GSHP	SHW	15,631.5	337.21	8,05.5
Split&DHW	Split	Split	Methane	7,350.0	1,496.9	2,716.9
MET&DHW	Methane	Split	Methane	7,350.0	1,151.25	3,354.8

Therefore, a payback period is expected for GSHP, as are significant savings over the long term. The emissions from GSHP plants are less than half of those of other plants, and it is more convenient to integrate GSHP and SHW plant than to use two separate plants.

Hereafter, the costs and benefits of each plant are analyzed over a period of 20 years. The problem is to decide which plant is cheaper, and with a house that already has a thermoregulation unit, it is convenient to switch to a GSHP plant. First, we compare the plants dedicated to thermoregulation: GSHP, METsplit, and Split.

Figure 5.8a shows the cumulative costs with an incentive of 36 % for all plants. The cumulative cost of GSHP appears to be stable over time due to the incentives that balance the maintenance costs. Whereas the cumulative cost of the METsplit and Split rapidly increases, the latter does not significantly exceed the cost of GSHP for 20 years.

Figure 5.8b shows the yearly difference in cost between METsplit and GSHP and between Split and GSHP, or, in other words, the savings of GSHP over the other plants. The GSHP plant achieves significant savings relative to the METsplit after 20 years, whereas there is no significant savings over the Split. The intersection between the trend lines and the time axis indicates the payback time of GSHP over the other two plants. The payback time of GSHP is 5 years relative to the METsplit, while the payback time is 17 years relative to the Split plant.

Figure 5.8c shows the yearly differences in cost between METsplit and GSHP and between Split and GSHP, expressed as compound interest. For a chosen plant, a compound interest is calculated over a number of years, considering the savings as *final capital* and the up-front cost difference between the chosen plant and the GSHP plant as *initial capital*. The compound interest represents the financial advantage of investing in GSHP rather than in another plant. GSHP provides significant savings over the METsplit, with the investment in GSHP rather than in METsplit providing the highest return of approximately 5 % over 10 years, with negative returns for periods of less than five years.

The Fig. 5.8d compares the equivalent emissions of GSHP, METsplit, and Split, which are constant unless the efficiency of the plants varies over the years.

The calculations are systematically applied in every site of the study region. The final output is GIS layers. The scheme for the calculation of the cumulative

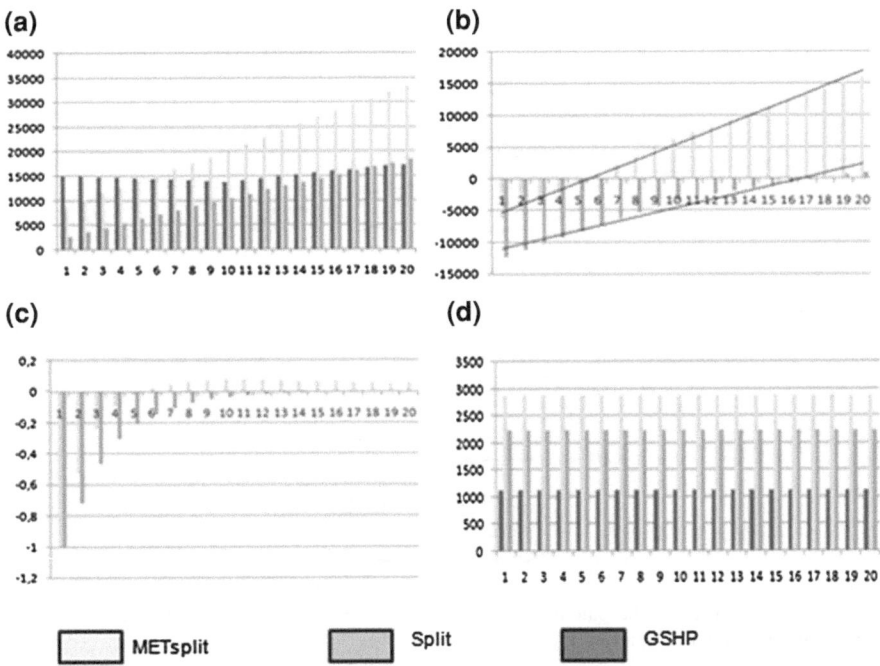

Fig. 5.8 Cost-benefit comparison of GSHP, METsplit, and split plants. **a** Absolute costs (€), **b** savings relative to GSHP (€), **c** interests (%), **d** emissions (CO_2 Kg/KWh)

savings layer and the payback years layer is presented in Fig. 5.9a, while the scheme for the calculation of compound interest layer is shown in Fig. 5.9b.

A cost-benefit comparison analysis of integrated plants providing thermoregulation and DHW is proposed here. The cost-benefits of GSHP&SHW, MET&DHW, and Split&DHW over a 20-year period are compared. The problem is to decide which plant is cheaper, and with a house that already has a thermoregulation and DHW system, it is convenient to switch to a GSHP&SHW plant. The calculation scheme in Fig. 5.9 is still valid.

Figure 5.10a shows the cumulative costs with an incentive of 36 % for all plants. The cumulative cost GSHP&SHW appears to be stable over time due to the incentives that balance the maintenance costs. Whereas the cumulative cost of the MET&DHW and Split&DHW rapidly increases, revealing that the integration of GSHP&SHW provides significant advantages over the other plants.

Figure 5.10b shows the yearly difference in cost between MET&DHW and GSHP&SHW, and between Split&DHW and GSHP&SHW, or, in other words, the savings of GSHP&SHW over the other plants. The GSHP&SHW achieves a significant saving relative to MET&DHW in 20 years. The intersection between the trend lines and the time axis indicates the payback time of GSHP&SHW over

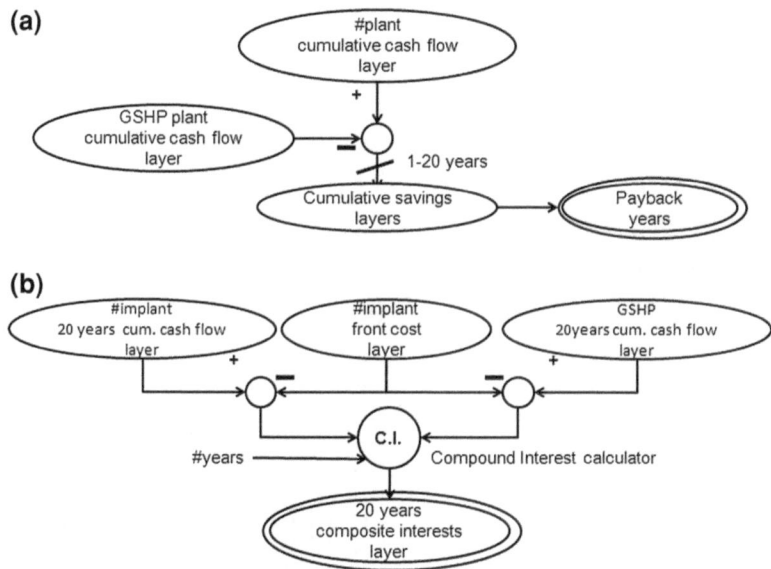

Fig. 5.9 **a** Procedure for the calculation of the cumulative savings layer. **b** Procedure for the calculation of compound interest layer

the other two plants. The payback time of GSHP&SHW is 5 years relative to the MET&DHW, whereas it is 7 years relative to the Split&DHW plant.

Figure 5.10c shows the yearly difference of costs between MET&DHW and GSHP&SHW, and between Split&DHW and GSHP&SHW, expressed as compound interest. The accrued interest by the GSHP&SHW investment is significant over the MET&DHW. Also, the curves show that the investment in the GSHP&SHW provides the highest return of approximately 5 % over 10 years, with negative returns for periods of less than five years.

The Fig. 5.10d compares the equivalent emissions of GSHP&SHW, MET&DHW, and Split&DHW, which are constant unless the efficiency of the plants varies over the years.

Figure 5.10 Cost-benefits comparison between GSHP&SHW, MET&DHW, and Split&DHW plants

The comparison between the systems dedicated to thermoregulation and those integrating thermoregulation and DHW shows that the economic advantage introduced by geothermal is stronger in the second group. We highlight this by comparing the savings in the following two cases:

- switching from MET&DHW to GSHP&SHW accrues a saving of about 20 k€ in 20 years (Fig. 5.10b), whereas switching from METsplit to GSHP accrues a saving of about 17 k€ (Fig. 5.8b);

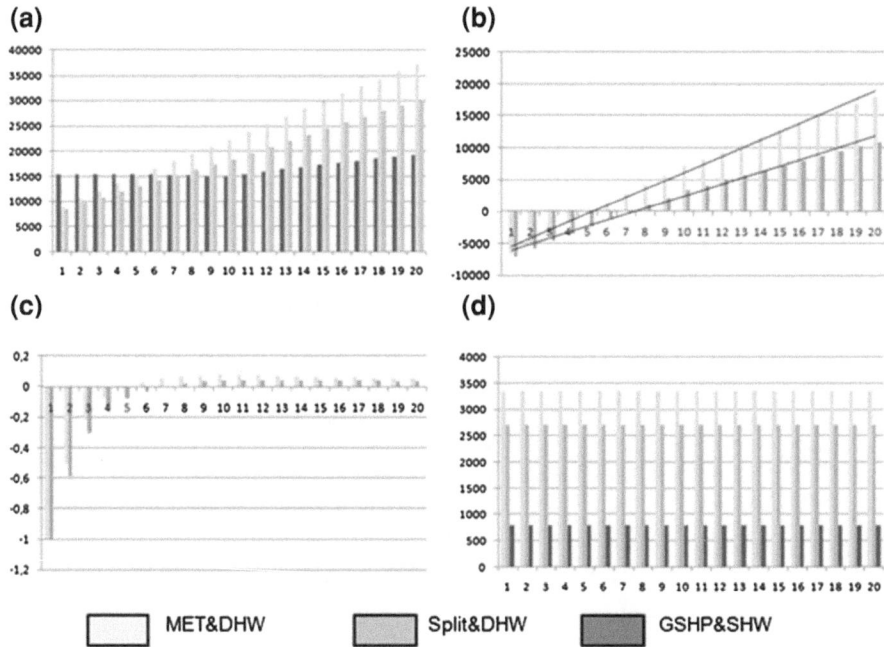

Fig. 5.10 Cost-benefits comparison between GSHP&SHW, MET&DHW, and Split&DHW plants. **a** Absolute costs (€), **b** savings relative to GSHP&SHW (€), **c** interests (%), **d** emissions (CO_2 Kg/KWh)

- switching from Split&DHW to GSHP&SHW accrues a saving of about 12 k€ in 20 years (Fig. 5.10b), whereas switching from Split to GSHP accrues a saving of about 2 k€ only (Fig. 5.8b).

References

European Community (2002) Directive 2002/91/EC—energy performance of buildings (EPBD). Directive 2002/91/EC
Lantschner N (2005) Casa Clima: Vivi in più. Edition Raetia, Bolzano
Santini, E, Elia S, Fasano G (2009) Caratterizzazione dei consumi energetici nazionali delle strutture ad uso ufficio, ENEA
Verein Deutscher Ingenieure (2001) Thermische Nutzung des Untergrundes–Blatt 2:erdgekoppelte Warmepumpenanlagen, Beuth Verlag, VDI-Richtlinie 4640

Chapter 6
Decision Analysis: Choosing the Right Plant

Abstract A series of decision problems concerning the choice of the optimal plant are defined and solved in each cell of the geographic grid. To this end, the cost-benefit analysis of the geothermal resource has been extended to the geographic dimension. The decision analysis is conducted on the basis of the decision criteria and priorities for different categories of stakeholders. The goal is to automate the decision problem solving, to build spatial decision-making models and represent them in map format.

This chapter applies decision analysis to solve the problem of choice of plant. The decision problem can be stated in the following way:

Problem 1: Decide Which Thermoregulation Plant Among GSHP, METsplit, and Split Offers the Greatest Overall Benefits for a Given Stakeholder.

The multi-criteria decision method has been used to solve this decision problem. The decision alternatives are the three alternative thermoregulation plants: GSHP, METsplit, and Split. The decision criteria are chosen among the features that compare the costs and benefits of the plants. For example, features such as payback time and compound interest both express the advantage of the GSHP plant relative to the other plants. The criteria taken into account for the decision process are as follows:

- Payback years of GSHP relative to METsplit;
- Payback years of GSHP relative to Split;
- Compound interest of GSHP relative to METsplit, over 20 years;
- Compound interest of GSHP relative to Split, over 20 years;
- Yearly emissions of METsplit;
- Yearly emissions of Split;
- Yearly emissions of GSHP.

For each site in the study area, the data matrix of Table 6.1 is available.

The MCDA requires also an input weight of each criterion. Three types of stakeholders have been identified: householders, investors, and administrations. The

Table 6.1 Decision criteria scheme for thermoregulation plants

Alternative plants	Criteria		
	Payback years of GSHP (relative)	Compound interest of GSHP (relative)	Emissions (absolute)
METsplit	Value	Value	Value
Split	Value	Value	Value
GSHP	0	0	Value

priorities between the expectations of every stakeholder are summarized in Table 6.2. It is assumed that the householders give the highest priority to the payback time, in second place to the emissions, and in third place to the compound interest. The weight of each criterion, which varies between 0 and 1, is assigned according to the priority. For each category of stakeholders, the sum of the weights equals 1.

Afterward, the decision problem is solved separately for each stakeholder. The results of the MCDA, systematically applied to each site of the study area, are reported and discussed in the atlas chapter of this book. Figure 7.9 shows the result of the decision process for the Fabriano site. The results show that the optimal decision from the point of view of the governmental agencies (Fig. 7.9c) and householders (Fig. 7.9a) is the GSHP, whereas the Split is the optimal choice for investors (Fig. 7.9b). However, the MCDA results are relative, indicating the optimal solution but not excluding the possibility that other solutions may have minor advantages.

The decision process is applied in a similar way to the integrated systems, thermoregulation, and DHW. The set of stakeholders is the same. The decision problem can be stated in the following way:

Problem 2: Decide Which Integrated Plant, GSHP&SHW, MET&DHW, or Split&DHW, Offers the Greatest Overall Benefits for a Given Stakeholder.

Given the GSHP&SHW as a reference plant, the following features are identified:

• Payback years of GSHP&SHW relative to MET&DHW;
• Payback years of GSHP&SHW relative to Split&DHW;
• Compound interest of GSHP&SHW relative to MET&DHW, over 20 years;
• Compound interest of GSHP&SHW relative to Split&DHW, over 20 years;
• Yearly emissions of MET&DHW;
• Yearly emissions of Split&DHW;
• Yearly emissions of GSHP&SHW.

Table 6.2 Priorities between stakeholders' expectations

Stakeholders	Priority		
	1	2	3
Householders	Payback	Emissions	Compound interest
Investors	Compound interest	Payback	Emissions
Administrations	Emissions	Payback	Compound interest

Table 6.3 Decision criteria scheme for thermoregulation and DHW plants

Alternative plants	Criteria		
	Payback years of GSHP&SHW (relative)	Compound interest of GSHP&SHW (relative)	Emissions (absolute)
MET&DHW	Value	Value	Value
Split&DHW	Value	Value	Value
GSHP&SHW	0	0	Value

For each site of the study area, the data matrix of Table 6.3 is available.

The set of priorities is the same as in the previous case. The calculations are systematically applied to each point in the region under study, and the results are reported in Fig. 7.9 of the atlas section. Figure 7.9a, b, and c shows that for each stakeholder, the decision is the system that integrates GSHP and SHW, and this system was chosen in the higher percentage of cases than the GSHP plants dedicated to thermoregulation (Fig. 7.9a, b, c).

Chapter 7
Regional Atlas Supporting the Decision-Making Process

Abstract In this chapter, a set of cartographic products has been gathered to support the decisions in the case study. Three different methods of analysis are highlighted, based on maps that represent different levels of support: (1) passive level, maps of individual features; (2) intermediate level, maps in which the decision criteria are represented; (3) active level, maps that express directly the solution to the decision problem. The collection constitutes a goal-oriented decision atlas dedicated to LTGE.

This chapter presents the results of calculations systematically applied to every site in the study region as maps. The aim is to set up an atlas supporting the spatial decision(s). The chapter is divided into three sections representing three different approaches, hierarchically organized from the simple description of the features to the description that provides solutions to given decision problems. The first section contains maps that describe the features of the cost of an individual system. In the second section, the features that compare costs between the plants are mapped. The third section provides maps that enable active support for decision problems. The maps are also discussed to highlight the economic implications of these decisions.

7.1 Mapping the Features

This section contains the maps that express the features of energy consumption and the cost of the plants, concerning the Problem 1.

Figure 7.1a shows the yearly energy needs of the GSHP plant for heating a 100 m^2 class E house. Areas with higher altitudes have higher energy needs due to the colder climate. The map is strongly correlated with the degree-day data of Fig. 4.1a. The up-front cost of the GSHP system across the region is shown in Fig. 7.1b. In the map, between the extreme values, there is a variation in up-front

A. Gemelli et al., *GIS to Support Cost-Effective Decisions on Renewable Sources*, SpringerBriefs in Applied Sciences and Technology, DOI: 10.1007/978-1-4471-5055-8_7, © The Author(s) 2013

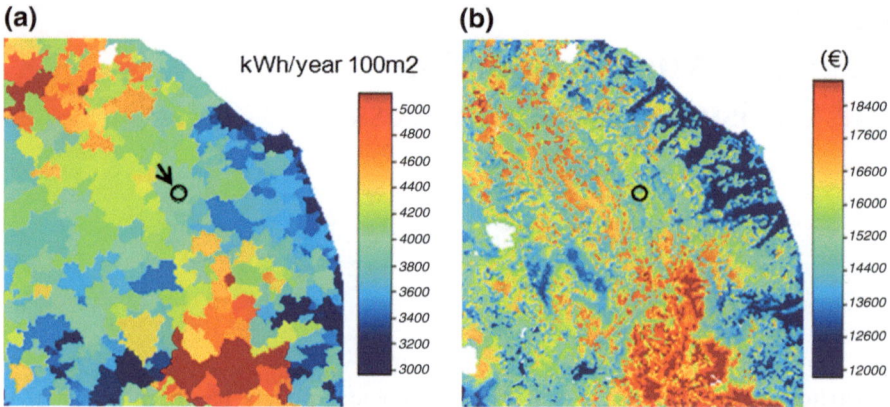

Fig. 7.1 a Yearly energy needs for heating a 100 m^2 class E house; the *arrow* indicates the location of Fabriano site. **b** Up-front cost of GSHP plant

costs of approximately 40 % due to the variability in energy needs and specific heat extraction. The average up-front cost of GSHP is 15,396.0 €. Across the region, the costs of METsplit and Split have been set at 7,350.0 € and 1,749.0 €, respectively.

The maintenance cost of the METsplit, Split, and GSHP plants is shown in Fig. 7.2a, b and c, respectively. The frequency of maintenance costs is compared in Fig. 7.2d. The GSHP system has the lowest average maintenance cost, € 485.2, while the maintenance cost of the systems METsplit and Split is 1,711.5 € and 1,058.2 €, respectively.

The CO_2 equivalent emissions of the METsplit, Split, and GSHP plants are shown in Fig. 7.3a, b, and c, respectively. The frequency of the cumulated costs for each type of plant is shown in Fig. 7.3d. The average cumulated cost of METsplit and Split is 33,860.2 € and 18,787.1 €, respectively. The average cumulated cost of GSHP is 17,778.5 € and then slightly below the Split plant.

The CO_2 equivalent emissions of the METsplit, Split, and GSHP plants are shown Fig. 7.4a, b, and c, respectively. For METsplit and Split, the average emissions are 2,939.8 kg/kWh year and 2,289.0 kg/kWh year, respectively (Fig. 7.4d). The GSHP has the lowest emission: only 1,156.8 kg/kWh year.

7.2 Mapping the Decision Criteria

In this section, the maps representing the decision criteria of Problem 2 are presented. The costs sustained in the first year for MET&DHW, Split&DHW and

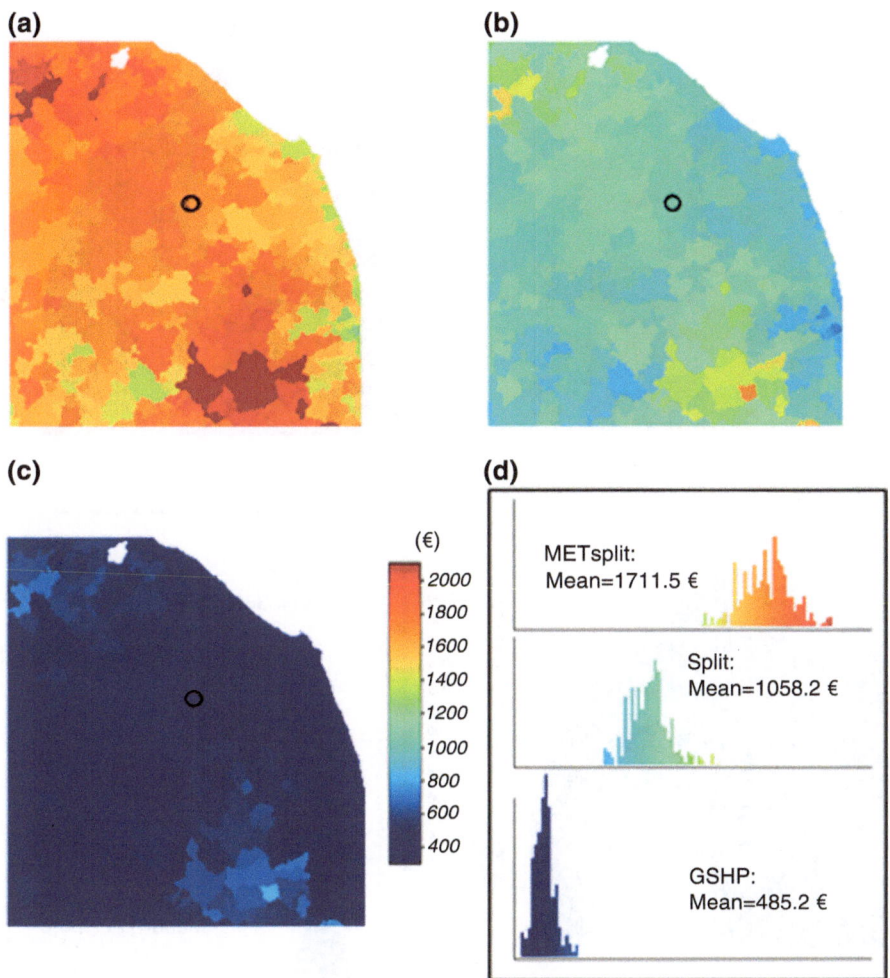

Fig. 7.2 a Yearly maintenance costs of the METsplit plant. **b** Yearly maintenance costs of the Split plant. **c** Yearly maintenance costs of the GSHP plant. **d** Relative frequency of costs related to plants

GSHP&SHW are shown in Fig. 7.5a, b, and c, respectively. It can be observed that the costs of MET&DHW and Split&DHW are similar to each other and 50 % lower than the cost of the GSHP&SHW, in the first year.

The payback time of the GSHP&SHW plant relative to an MET&DHW plant is shown in Fig. 7.6a. The payback period is the period of operation required for the GSHP&SHW to become advantageous over the MET&DHW. Similarly, the

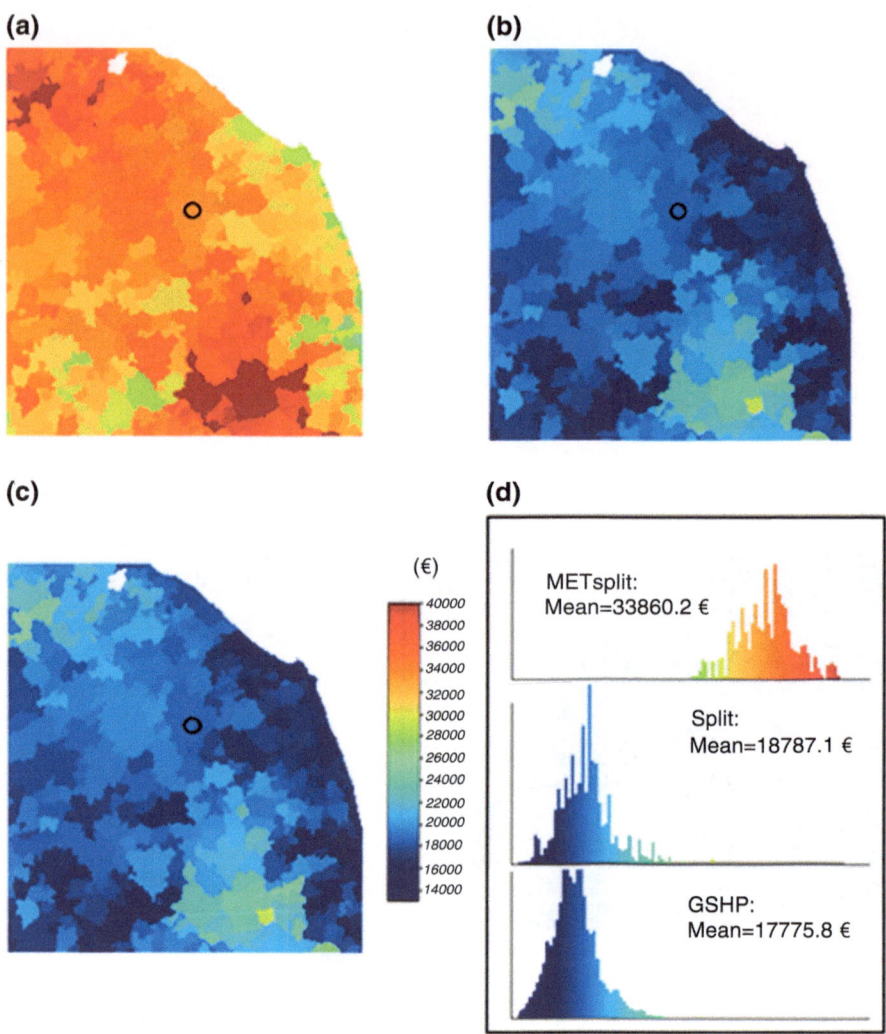

Fig. 7.3 a Cumulated costs of the METsplit plant. **b** Cumulated costs of the Split plant. **c** Cumulated costs of the GSHP plant. **d** Relative frequency of costs related to plants

payback time of the GSHP&SHW plant relative to a Split&DHW plant is shown in Fig. 7.6b. It is possible to observe that in the second case, the payback time is longer, namely the conversion from MET&DHW to GSHP&SHW is amortized in fewer years than it is the conversion from Split&DHW to GSHP&SHW. The payback period is taken as the first decision criterion of Problem 2.

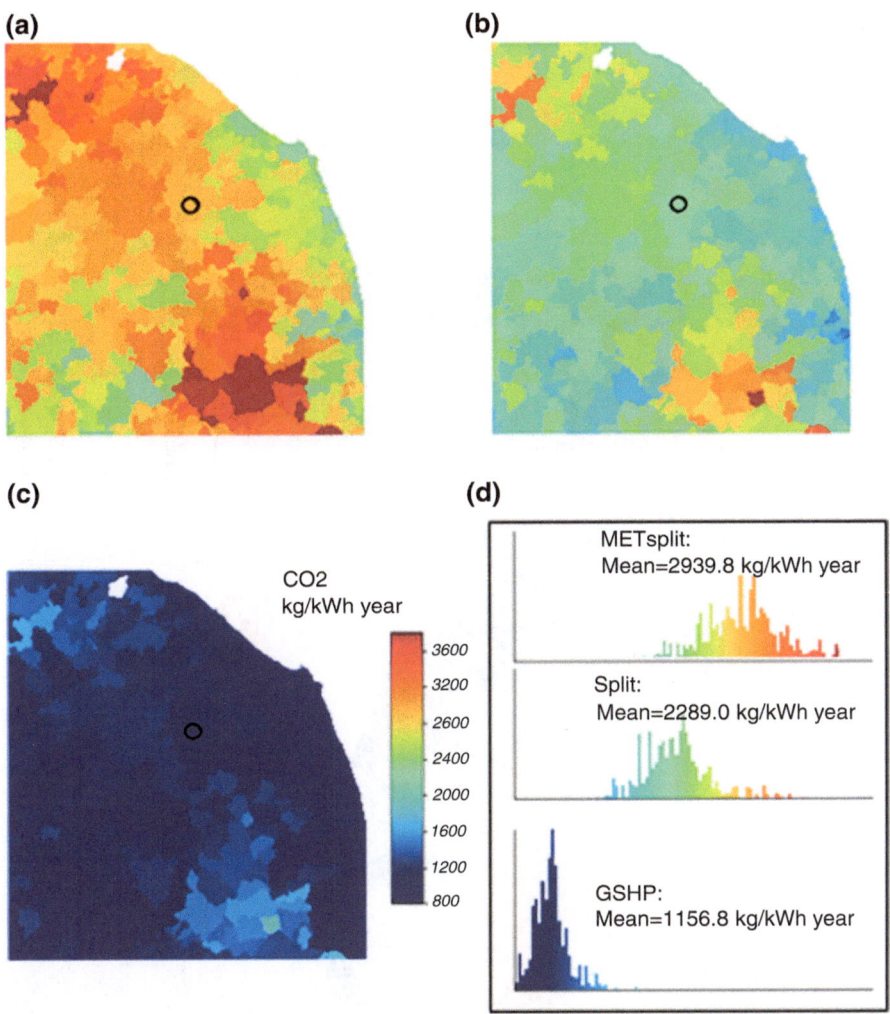

Fig. 7.4 a Emissions of the METsplit plant. **b** Emissions of the Split plant. **c** Emissions of the GSHP plant. **d** Relative frequency of emissions related to plants

The savings accumulated over 20 years by the GSHP&SHW system compared with the MET&DHW are expressed as compound interest in the map of Fig. 7.7a. The compound interest of GSHP&SHW over Split&DHW (Fig. 7.7b) is considerably lower across the region. The conversion from MET&DHW to GSHP&SHW would be a financially more interesting investment than the conversion from Split&DHW to GSHP&SHW. In all cases, GSHP&SHW is the most convenient

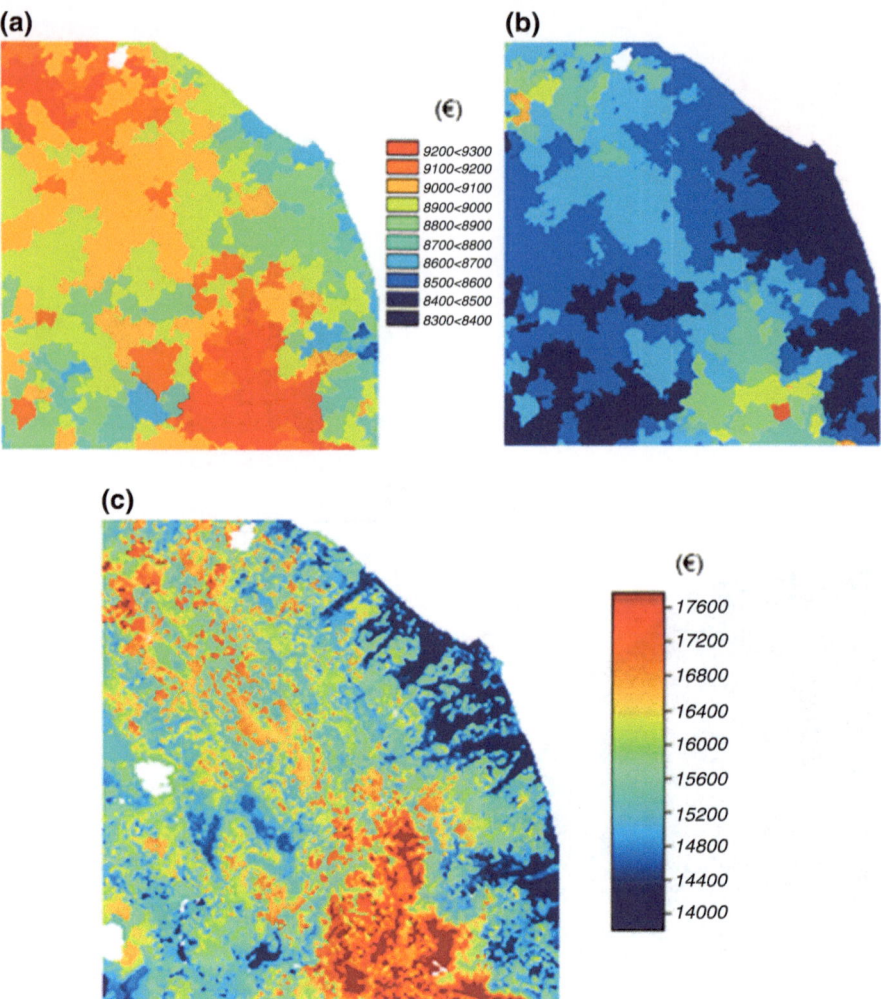

Fig. 7.5 **a** Up-front cost of MET&DHW plant. **b** Up-front cost of Split&DHW plant. **c** Up-front cost of GSHP&SHW plant

form of investment. The feature of compound interest is taken as the second decision criterion in Problem 2.

The CO_2 equivalent emissions for MET&DHW, Split&DHW, and GSHP&SHW are shown in Fig. 7.8a, b, and c, respectively. Note that the CO_2 equivalent emissions of GSHP&SHW are significantly lower than those of the other two types of plants. The feature of equivalent emissions is taken as the third decision criterion in Problem 2.

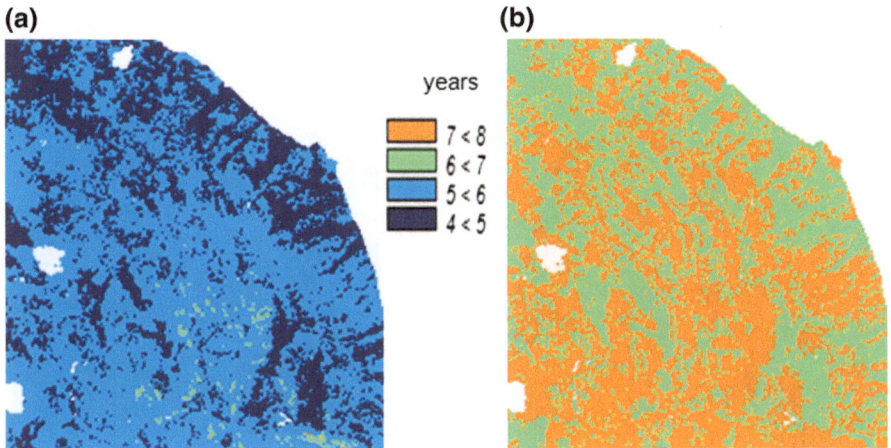

Fig. 7.6 a Payback years of GSHP&SHW over an MET&DHW plant. **b** Payback years of GSHP&SHW over a Split&DHW plant

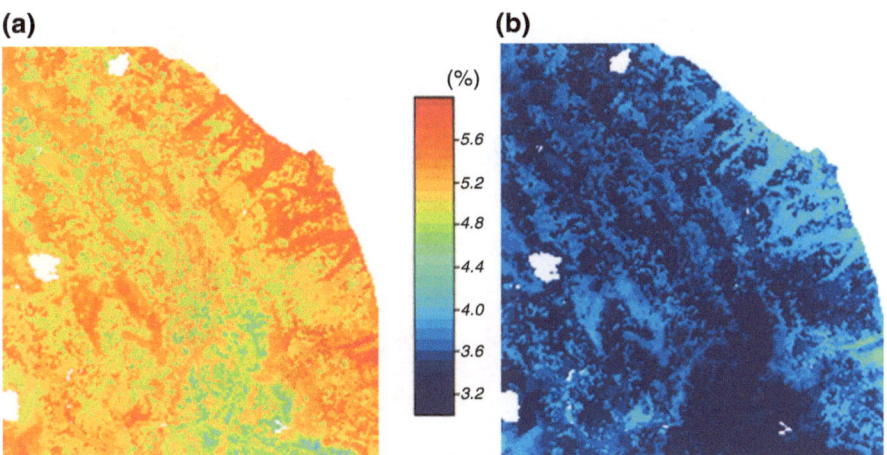

Fig. 7.7 a Compound interest of GSHP&SHW over an MET&DHW plant. **b** Compound interest of GSHP&SHW over a Split&DHW plant

7.3 Multi-Criteria Evaluation Analysis

The previous sections of this chapter only showed maps expressing individual decision criteria. Although each criterion allows for statements to be made about the advantage of one type of plant over another, the comparative analysis of all the

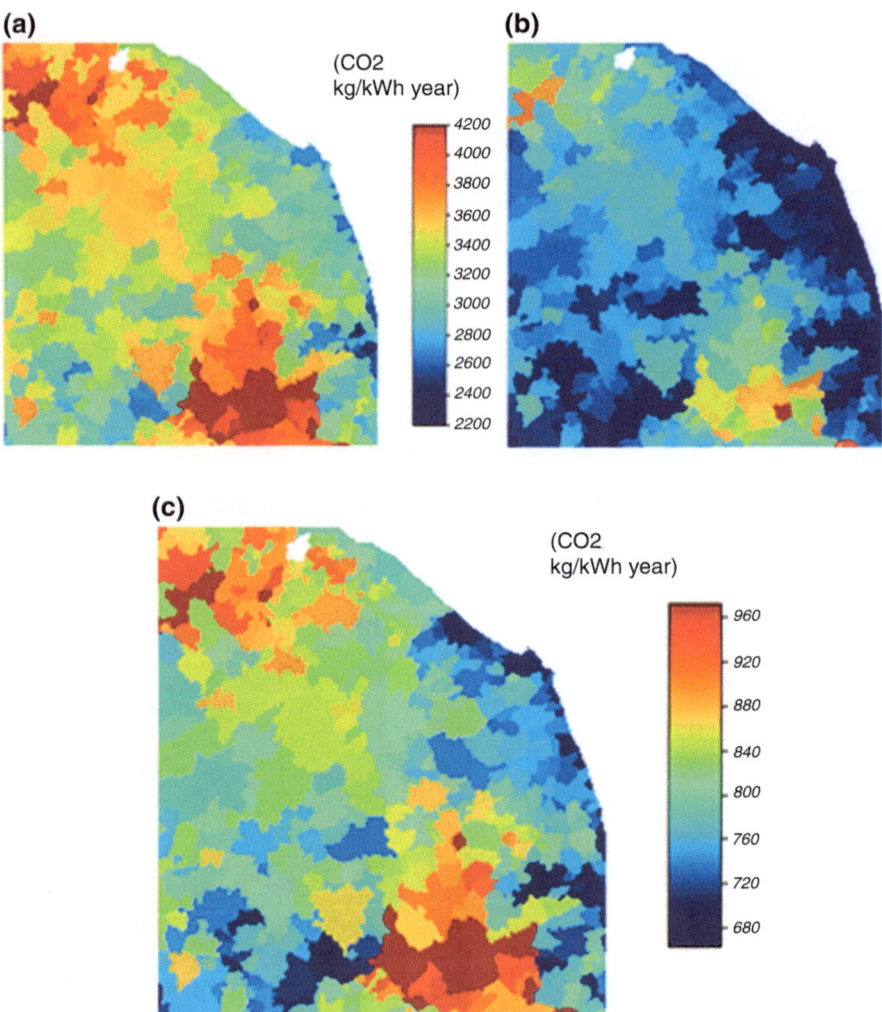

Fig. 7.8 **a** Emissions of the MET&DHW plant. **b** Emissions of the Split&DHW plant. **c** Emissions of the GSHP&SHW plant

criteria used to find the optimum plant would be a difficult mental process. This is due of the large amount of information that must be evaluated. For this reason, the maps of the individual features are only passive support for the decision. In contrast, this section presents a set of maps that provide active support for the decision problem, as in each map, the decision criteria are combined into one single feature expressing the solution to the decision problem. These maps are produced by MCDA. Each map corresponds to a decision problem, and each pixel is a response to the decision problem in a geographic site.

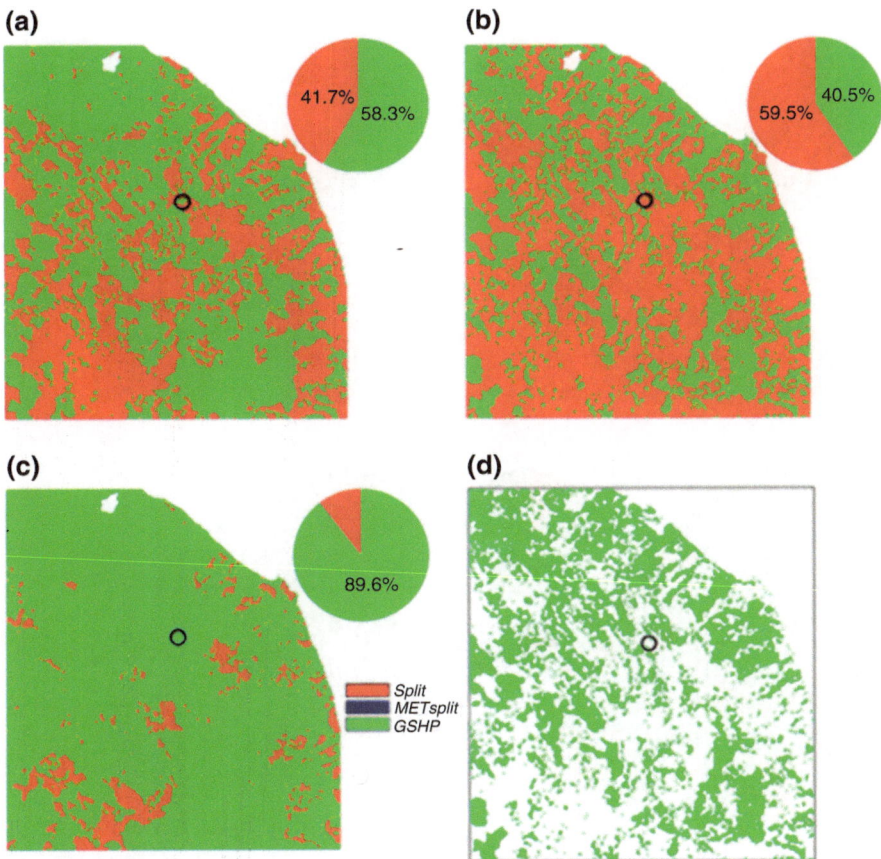

Fig. 7.9 Solution to the thermoregulation plant's decision problem. **a** Householders' decision. **b** Investors' decision. **c** Administration decision. **d** Common decision

Fig. 7.9 presents the solution to Problem 1, namely the decision between the alternative plants METsplit, Split, and GSHP, from the perspective of the three stakeholders: householders, investors, and administrations (Fig. 7.9a, b, c). Different stakeholders make different decisions: for the householders, the investors, and the administration, the GSHP plant is optimal in 58.3, 53.3, and 89.6 % of sites, respectively. Note that the alternative solution METsplit is not chosen in any site. The decision is in favor of GSHP for all stakeholders across 38.1 % of the region, as shown in, Fig. 7.9d.

Fig. 7.10 presents the solution to Problem 2, namely the decision between the alternative plants MET&DHW, Split&DHW, and GSHP&SHW from the perspective of the three stakeholders: householders, investors, and administration

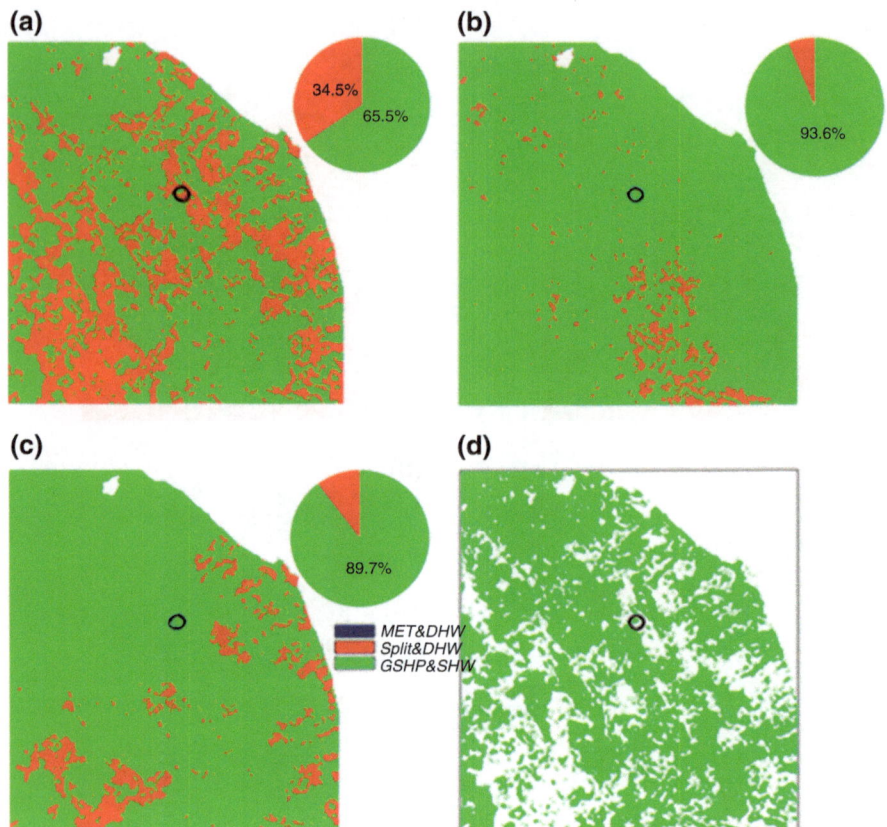

Fig. 7.10 Solution to the thermoregulation and DHW plant's decision problem. **a** Householders' decision. **b** Investors' decision. **c** Administration decision. **d** Common decision

(Fig. 7.10a, b, c). For the householders, the investors, and the administration, the GSHP&SHW plant is optimal in 65.5, 93.6, and 89.7 % of sites, respectively. The alternative solution MET&DHW is not chosen in any site. The sites where the decision is in favor of GSHP&SHW for all stakeholders constitute 62.2 % of the region, Fig. 7.10d.

All the maps shown above in this section are based on a scenario with government incentives of 36 %, whereas Fig. 7.11 shows the common decision in a

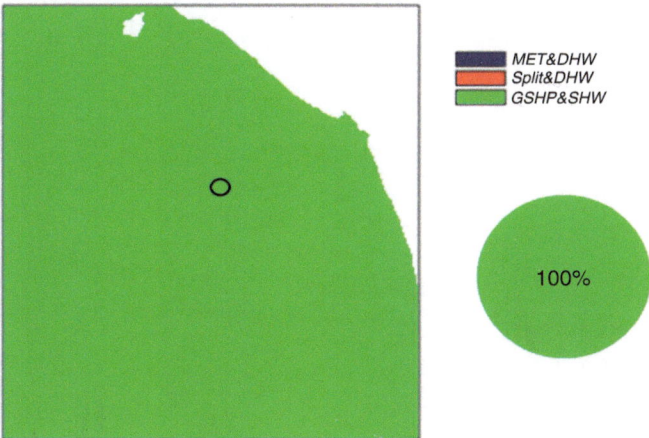

Fig. 7.11 Higher incentive scenario solution to thermoregulation and DHW plant's decision problem

scenario with incentives of 55 %. In latter case, the decision is ubiquitously favorable to the solution that integrates the ground-source heat pump with the solar water heating.

Chapter 8
Conclusive Remarks

Abstract This chapter highlights the capabilities of a decision support system implemented for low temperature geothermal energy. Two main results were obtained: (a) to expand the architecture of a GIS to support the decision-making process through instruments based on automated reasoning, sustaining geocomputational approach to geographic data analysis; (b) to extract new and valid conclusions concerning of decision-making process on LTGE, contributing to structure this process and bring it to a high level of generality. A benchmarking of LTGE and other alternative plants for thermoregulation has been conducted, which shows undoubtedly the advantages offered by LTGE.

In this work, a study of geothermal resources has been performed on a wide region using GIS technology. The GIS allows the application of complex analytical procedures in a systematic manner in each site in the region, automating a complex decision-making process. For its multi-disciplinary approach, this case study represents an original application by contributing (a) new knowledge on geothermal energy planning and (b) advancements in the informative system design for geospatial decision support. These two aspects of the work are briefly summarized in this section.

8.1 Contribution to Informative System Design for Geospatial Decision Support

On a real case study concerning the low temperature geothermal energy, a spatial decision support project has been designed and implemented based on open source technology and publicly available data. The typical GIS architecture has been extended with data mining tools. A procedure based on the algorithms goal-oriented feature ranking and AdaBoost allows for the classification of the rocks forming the heat storage reservoir on the basis of MODIS satellite imagery. The information

A. Gemelli et al., *GIS to Support Cost-Effective Decisions on Renewable Sources*,   83
SpringerBriefs in Applied Sciences and Technology,
DOI: 10.1007/978-1-4471-5055-8_8, © The Author(s) 2013

system allows for the selection of the data and optimal algorithms for achieving analytical goals.

The workflow of the decision-making process is territorially exportable, applicable to other energy sources and open to any georeferenced information of economic nature. In a unique processing flow, the analysis of natural and technical factors and their economic effects are comprised, creating an original synthesis of a complex information environment. The output of the process consists of a model of the geothermal resource and the decision support models for different stakeholders.

8.2 Contribution to the Knowledge of Geothermal Resources

The decisions calculated with multi-criteria decision method indicate that domestic thermoregulation with geothermal energy is always more cost effective than methane-based thermoregulation, but GSHP is not always advantageous compared to split-based thermoregulation plants, depending on the site. When thermoregulation is associated with the production of domestic hot water, the integrated solution GSHP and solar hot water is highly competitive compared to any other alternative. The incentives have a significant effect on the cost benefit of the plant, which can radically change the scenario in favor of one solution or the other. Different stakeholders perceive differently the benefits of geothermal plants. Generally, the LTGE is more advantageous to the government, emphasizing the reduction in polluting emissions, and advantageous to a variable extent to investors and householders who are more interested in the economic aspect of LTGE.

The study shows two significant aspects in the field of geothermal energy: (a) the comparative analysis of different types of plants is crucial to decision making on LTGE, (b) it is always possible to improve the benefits from geothermal resource by integrating solar energy plants.

8.3 Future Works

This decision support system is open to further improvements. During data collection, several important features could be taken into account to further improve the cost-benefit analysis, especially features regarding the environmental impacts of the plants, technical variants, and daily and seasonal variations of the resource potential. Furthermore, it would be useful to consider the value added to plants by intelligent control similar to the smart grid concept applied to the electrical power grid. The performance of MCDA in problem solving can be improved with a detailed study on the interests of stakeholders leading to an objective weighting of the decision criteria.